The Rand McNally Library of Astronomical Atlases
for Amateur and Professional Observers
Series Editors Garry Hunt and Patrick Moore

The Moon

Patrick Moore

Maps by Charles A. Cross
Foreword by Professor Archie E. Roy

Published in Association with
the Royal Astronomical Society

Rand McNally & Company
New York · Chicago · San Francisco

The Moon
© Mitchell Beazley Publishers 1981

ISBN 528-81541-5
Library of Congress Catalog Number 80-53881

The Moon was edited and designed by
Mitchell Beazley Publishers,
Mill House, 87–89 Shaftesbury Avenue,
London W1V 7AD

Phototypeset by Servis Filmsetting Ltd.
Origination by Adroit Photo Litho Ltd.
Printed in the United States

Editor Gilead Cooper
Designer Wolfgang Mezger
Editorial Assistant Charlotte Kennedy
Picture Research Elizabeth Ogilvie

Executive Editor Lawrence Clarke
Art Manager John Ridgeway
Production Manager Barry Baker

The units and notation used throughout this book are based on the Système International des unités (SI units), which is currently being introduced universally for scientific and educational purposes. There are seven "base" units in the system: the *meter* (m), the *kilogram* (kg), the *second* (s), the *ampere* (A), the *kelvin* (K), the *mole* (mol) and the *candelo* (cd). Other quantities are expressed in units derived from the base units; thus, for example, the unit of force the newton (N) is defined as the force required to give a mass of one kilogram an acceleration of one meter per second squared ($kg\,m\,s^{-2}$).

Some branches of science continue to adhere to a few of the older units, and in one case an editorial concession has had to be made to existing scientific usage: the SI unit of magnetism, the tesla, has been dropped in favor of the more common unit, the gauss. One tesla is equal to 10,000 gauss.

For very large and very small numbers, "index notation" has been adopted, so that where appropriate numbers are written as powers of ten. For example, 1,000,000 may be written as 10^6, and 3,500,000 as 3.5×10^6. Numbers smaller than one are indicated by negative powers: thus 0.00035 is written as 3.5×10^{-4}. In addition a variety of prefixes is used to denote certain multiples of units (*see* Table 1). Table 2 gives the SI equivalents of common imperial units while Table 3 lists a selection of astronomical constants.

Table 1: SI prefixes

Factor	Name	Prefix Symbol
10^{18}	exa	E
10^{15}	peta	P
10^{12}	tera	T
10^9	giga	G
10^6	mega	M
10^3	kilo	k
10^2	hecto	h
10^1	deca	da
10^{-1}	deci	d
10^{-2}	centi	c
10^{-3}	milli	m
10^{-6}	micro	μ
10^{-9}	nano	n
10^{-12}	pico	p
10^{-15}	femto	f
10^{-18}	atto	a

Table 2: SI conversion factors

Length	
1 in	25.4 mm
1 mile	1.609344 km
Volume	
1 imperial gal	4.54609 cm³
1 US gal	3.78533 liters
Velocity	
1 ft/s	$0.3048\,m\,s^{-1}$
1 mile/h	$0.44704\,m\,s^{-1}$
Mass	
1 lb	0.45359237 kg
Force	
1 pdl	0.138255 N
Energy (work, heat)	
1 cal	4.1868 J
Power	
1 hp	745.700 W
Temperature	
°C	= kelvins − 273.15
°F	= $\frac{9}{5}$ (°C) + 32

Table 3: Astronomical and physical constants

Astronomical unit (A.U.)	1.4959787×10^8 km
Light-year (l.y.)	9.4607×10^{12} km = 63,240 A.U. = 0.306660 pc
Parsec (p.c.)	30.857×10^{12} km = 206,265 A.U. = 3.2616 l.y.
Length of the year	
Tropical (equinox to equinox)	$365^d.24219$
Sidereal (fixed star to fixed star)	365.25636
Anomalistic (apse to apse)	365.25964
Eclipse (Moon's node to Moon's node)	346.62003
Length of the month	
Tropical (equinox to equinox)	$27^d.32158$
Sidereal (fixed star to fixed star)	27.32166
Anomalistic (apse to apse)	27.55455
Draconic (node to node)	27.21222
Synodic (New Moon to New Moon)	29.53059
Length of day	
Mean solar day	$24^h03^m56^s.555 = 1^d.0027379\,1$ mean solar time
Mean sidereal day	$23^h56^m04^s.091 = 0^d.99726957$ mean solar time
Earth's sidereal rotation	$23^h56^m04^s.099 = 0^d.99726966$ mean solar time
Speed of light in vacuo (c)	$2.99792458 \times 10^5\,km\,s^{-1}$
Constant of gravitation	$6.672 \times 10^{-11}\,kg^{-1}\,m^3\,s^{-2}$
Charge on the electron (e)	$= 1.602$ coulomb
Planck's constant (h)	$= 6.624 \times 10^{-34}$ J s
Solar radiation	
Solar constant	$1.39 \times 10^3\,J\,m^{-2}\,s^{-1}$
Radiation emitted	$390 \times 10^{26}\,J\,s^{-1}$
Visual absolute magnitude (M_v)	+ 4.79
Effective temperature	5,800 K

Contents

Foreword

Astronomy is the oldest of the sciences, born of the fact that man evolved on a planet from which he could see the sky. For 50,000 years he has had intelligence enough to study the heavens and attempt to understand what he saw by day and by night. And there is no doubt that what he saw and deduced from his observations has had extraordinary effects upon his life in many lands.

If our civilization had developed in the way it has on a planet whose skies were eternally cloud-covered (which is doubtful), we would have believed, up to some forty years ago, that the Earth *was* the universe. Only with the advent of large radar dishes, high-flying aircraft and rockets would the shattering fact have emerged that above the opaque cloud-layer lay a seemingly boundless universe.

Modern western civilization had been greatly influenced by Copernicus, Kepler, Galileo and Newton, all watchers of the skies. Our belief in a rational universe capable of being understood and our scientific and technological civilization spring from the cyclic behaviour of Sun, Moon, planets and stars and Newton's ability to explain so much of that behaviour by his law of gravitation and his three laws of motion. Timekeeping, navigation, geodesy, dynamics, religious and philosophical systems, cosmology and relativity and many other activities and interests of man have been directly affected by our study of the heavens.

There have been three astronomical revolutions. The first—the serious, naked-eye study of the heavens—lasted a long time—at least five millennia—and ended when Galileo began his systematic telescopic study of the sky in AD 1610. That second revolution, in which the camera and the spectroscope played their part, was brought to a climax by the enormous amount of information gathered by telescopes such as the 200-inch Hale telescope at Mt Palomar. On October 4, 1957, with the orbiting of Sputnik I, the third astronomical revolution began. Not only can we now place instruments in orbit above the Earth's atmosphere, obtaining access to the entire electromagnetic spectrum, but we can send spacecraft such as the Mariners, Pioneers and Voyagers to other planets in our Solar System. The flood of astronomical information has become a torrent, sweeping away many of our former ideas about the universe.

The time therefore seemed ripe for a series of atlases designed to take stock of this flood of new information and the new under-standing it has brought us of the nature of the universe. Each atlas in the series has been written by an author chosen by Mitchell Beazley Publishers so that the text will provide the most up-to-date assessment of the celestial body studied, together with explanatory diagrams and the most modern pictures. Each author's text has then been carefully checked and authenticated by an acknowledged expert in the subject chosen by the Royal Astronomical Society's Education Committee. The final text of each book should therefore truly convey our present-day knowledge of the subject and remain a definitive work for many years to come.

Archie E. Roy
BSc, PhD, FRAS, FRSE, FBIS
Titular Professor of Astronomy in the
University of Glasgow
Chairman of the Education Committee of the
Royal Astronomical Society

Introduction

The Moon is our companion in space. It keeps together with us as we journey round the Sun and is much the closest of all natural astronomical bodies. Obviously, therefore, we tend instinctively to regard it as important.

Observation of the Moon was important to the earliest civilizations as a means of measuring time and keeping track of the seasons, and this activity became closely associated with religion. Lunar worship was widespread, and the Moon was regarded either as a god, or as the dwelling-place of a god. The Egyptian god Thoth, patron of learning and the arts, was associated with the Moon, as was the deity Khons, who cured the sick and exorcised spirits. The Babylonian Moon-god Sin was the lord of the calendar as well as of wisdom. The Greeks worshipped the Moon as Selene, sister of Helios, the Sun-god, and the Romans later identified Selene with their own Moon-goddess, Luna. More than a dozen other names for a female Moon-deity can be found in classical literature.

The Moon has also held an important place in astrology, in which it has been associated particularly with the soul and the subconscious. At various times it has also been thought to influence or even cause madness. In Christian iconography the Moon has often been linked with the Virgin Mary.

At an early stage it was recognized that the Moon is the main cause of the tides. However, some of the old ideas were very strange indeed. The Greek philosopher Xenophanes, for example, who died in or around 478 BC, wrote that "there are many suns and moons according to the regions, divisions and zones of the Earth"; he regarded the Earth as flat, with its surface touching the air and its underside extending without limit.

Gradually, the concept of the Moon as a body moving around a spherical Earth began to gain support, and around 270 BC Aristarchus of Samos made a very reasonable estimate of its distance. There is also evidence that Aristarchus believed the Earth to be in orbit around the Sun, thereby anticipating Copernicus by nineteen centuries. In his essay *On the Face of the Orb of the Moon*, written about AD 80, Plutarch maintained that the Moon must be "earthy", with mountains and ravines. Yet the real character of the Moon remained uncertain, and as recently as the early nineteenth century the great astronomer William Herschel, the discoverer of the planet Uranus, was still insisting that the habitability of the Moon was "an absolute certainty".

The scientific study of the Moon has been particularly concerned with the problem of mapping the Moon's surface, a study known as "selenography". In the years before the war, the best lunar maps were of amateur constuction, and valuable conclusions were drawn from them. The situation now is very different. Cartography has to all intents and purposes been completed, and no Earth-based results can show as much fine detail as those from the space-probes. Yet, perhaps surprisingly, there is still scope for amateur work in more specialized fields, notably the detection of the short-lived glows and obscurations known as Transient Lunar Phenomena or TLP (*see* page 20).

This atlas will be of particular value to the practical observer of the Moon. It includes two different types of map, each with a special purpose: the first type shows the near side of the Moon in the form most convenient to the observer, using an orthographic projection, which depicts the hemisphere as it appears from Earth. The second type uses Mercator and polar stereographic projections to provide coverage of the entire globe without the foreshortening that occurs in an Earth-based view; these maps are ideal for comparison with the accompanying photographs from space.

There is endless enjoyment to be gained simply from looking at the Moon, even with a telescope of modest aperture. The view is different from one night to the next; there is an ever-changing panorama of mountains, craters, valleys and peaks, and despite the unmanned probes and the Apollo missions the Moon has lost none of its romance or its fascination.

1. The Moon
Composite of photographs taken at the first and third quarter. Lick Observatory, California.

2. The Goddess Luna
Roman sculpture, first century AD.

3. Symbol of the Moon

Origin and Status

Although relatively insignificant in terms of the Solar System, the Earth's Moon is outstandingly large and massive in relation to its primary. It has a diameter of 3,476 km, over a quarter the size of the Earth's diameter, and a mass of 7.3483×10^{22} kg, as compared with a figure roughly 81 times larger for the Earth. There are five satellites that are larger than the Moon—three in Jupiter's system, one in Saturn's and one in Neptune's—but all these move around giant planets which are several hundred times more massive than their respective satellites. The Earth–Moon system has sometimes been described as a "double planet" to emphasize the importance of the Moon, although the term carries with it certain assumptions about the origin of the two bodies which may or may not be valid.

Origin

Various suggestions have been put forward concerning the origin of the Moon, although none of them has been confirmed. Existing theories may be grouped into three main categories, according to which basic mechanism they invoke: "fission", "capture" or "binary accretion".

The "fission hypothesis" was first proposed by G. H. Darwin in 1878. Darwin started by assuming that the Earth and the Moon were originally one body, and that the Moon was thrown off as a fluid mass when rapid rotation caused the combined body to become unstable; it became first pear shaped, then dumb-bell shaped, with one "bell" much larger than the other. Eventually there was a break in the narrow neck of the body, and the Moon moved away, settling into a stable orbit. It was even suggested that the Pacific Ocean marked the scar left by the event, although the Pacific is far too shallow to account for a body as large as the Moon. Moreover, it has been calculated that the Earth would have needed to spin extremely rapidly, with a period shorter than about 3 hr, in order for there to be sufficient energy to separate the Moon from the Earth in the manner suggested; special reasons would be needed to explain why the Earth should have been spinning so rapidly.

A more popular idea is that the Moon used to be an independent planet formed elsewhere in the Solar System and was subsequently captured during a close approach to the Earth, so that the two have remained together ever since. Admittedly the probability of such an event occurring is extremely small, since the capture of one body by the gravitational field of another can only happen under certain conditions. The approaching body would have to be decelerated (or else it would simply escape the Earth's gravitational pull), but the mechanisms for slowing it down, such as tidal interactions between the two bodies, are not able to account for a very dramatic change in velocity. Thus a limit is imposed on the kind of orbit that might have been followed by the Moon before capture: if the orbit was too eccentric, then the Moon would be travelling too fast when it approached the Earth. Perhaps the Moon collided with smaller objects already in orbit about the Earth, and these contributed to its deceleration; such a mechanism might make the capture hypothesis a more likely explanation.

The third type of theory is based on the idea of "binary accretion", proposing that the Earth and the Moon were formed at the same time, out of the same material and in the same region. There is general agreement that the planets were produced from material in a "solar nebula", a cloud of material associated with the primordial Sun, and analyses of the rocks brought back from the Moon by space missions have shown it to be of approximately the same age as the Earth: between 4,500 and 4,600 million years. However, the Earth and the Moon are very different in density and chemical composition, and this makes it difficult to accept the idea of a common origin in close proximity. There are also dynamical arguments against the double planet hypothesis.

There is, finally, a fourth model known as the "precipitation hypothesis", which combines elements of both the fission and the binary accretion theories. According to this view, energy released

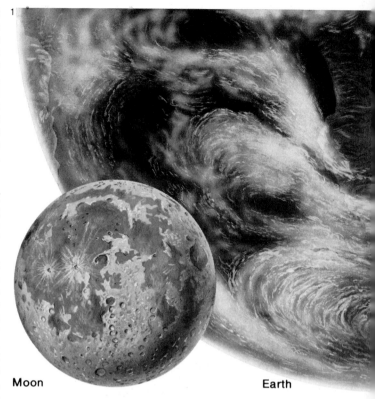

Moon   Earth

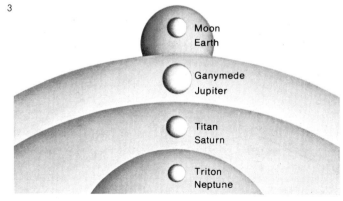

1. Scale of the Moon
The Moon's diameter (3,476 km) is more than a quarter that of the Earth; the mass ratio is 1:81.

2. Earth–Moon distance
The Moon orbits the Earth at a mean distance of 384,392 km, less than 10 times the circumference of the Earth.

3. Satellites and their primaries
Relative to its primary (Earth), the Moon is unusually large and massive. By comparison Triton has only about one-eighth the diameter of Neptune, with a mass ratio of 1:750; for the largest satellites of Saturn (Titan) and Jupiter (Ganymede) the ratio is even smaller.

4. Origin of the Moon
According to the "fission theory" (**A**) the Moon once formed part of the Earth, and broke away as a result of the Earth's rapid rotation. The "precipitation hypothesis" (**B**) suggests that the Moon was built up by accretion of particles spun off the Earth soon after the Earth itself was formed. Another possibility is that the Moon was formed together with the Earth and in the same manner. The "capture hypothesis" (**C**) proposes that the Moon began as an independent body, formed elsewhere in the Solar System, and that it was captured by the gravitational field of the Earth during a close approach to the planet.

Moon
Earth

Ganymede
Jupiter

Titan
Saturn

Triton
Neptune

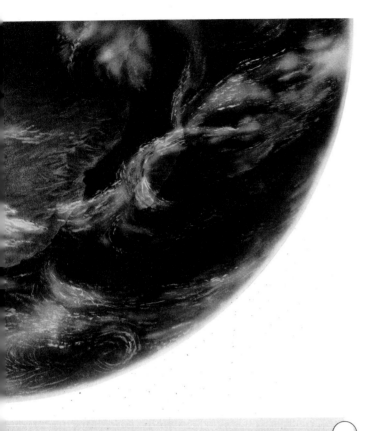

during the formation of the Earth heated up part of the material, forming a hot, dense atmosphere of metal and oxide vapors around the planet. These hot gases were then spun out from the Earth's equator, gradually cooling and condensing into grains of dust which eventually accreted to produce the Moon. This version of events has the advantage of explaining the Moon's lack of a large dense core corresponding to that of the Earth.

Atmosphere and surface conditions

The Moon is a hostile world for one main reason: its lack of atmosphere. Because of its small size and low mass, the Moon's gravitational field is relatively weak; in fact the acceleration due to gravity on its surface is only one-sixth that on the surface of the Earth. Consequently, the Moon's "escape velocity" is low—2.37 km s^{-1}—which is not high enough for it to retain an appreciable atmosphere. This has been confirmed both by Earth-based astronomers and by the Apollo astronauts.

One method of measuring the Moon's atmosphere depends on careful observations of bright stars as they are hidden or "occulted" by the Moon moving across the sky. If the Moon had an atmosphere, the stars would appear to flicker and fade shortly before they disappeared behind its limb; in fact they appear to snap out instantaneously, in contrast with occultations of stars by, for example, Venus, which does have an atmosphere. (This comparison can readily be checked by amateurs.) The accuracy of the experiment may be improved by measuring the degree of polarization (*see* Glossary) that occurs as a result of the scattering of light by any gaseous molecules around the Moon, and even more accurate measurements can be made by observing the occultation of radio sources. From such experiments it can be calculated that the density of the Moon's atmosphere must be 10^{-14} times less than that of the Earth, making it virtually indistinguishable from a hard vacuum. However, an experiment set up in December 1972 by the astronauts of Apollo 17 revealed the presence of an excessively tenuous atmosphere on the Moon's surface; gases detected included hydrogen, helium, neon and argon, but the density is so low that the molecules would practically never collide with one another.

Thus, whereas the surface of the Earth has been subject to a wide variety of climatic conditions which have influenced its geological history, the Moon's surface has been unaffected by these weathering processes. Similarly, the absence of free water on the Moon means that another of the principal causes of erosion on the Earth is lacking, and it would therefore be expected that the Moon's surface should differ substantially from that of the Earth.

Other forces, on the other hand, play a considerably greater part in shaping the surface of the Moon. There is, for example, a steady bombardment by particles and fragments of rock, ranging in size from large boulders (which are fairly rare) down to particles of dust. On the Earth most of these particles burn up in the atmosphere before reaching the ground and may be seen as meteors or "shooting stars", but on the Moon they strike the surface at high velocities, eroding the rock and producing craters of various sizes. The Moon is also bombarded by the "solar wind", which is emitted by the Sun and which consists mainly of protons and electrons; since these particles are electrically charged, they tend to be deflected by the Earth's magnetic field, whereas the Moon's magnetic field (*see* page 21) is too weak to provide the same protective influence.

Comparative quantities for Earth and Moon

	Equatorial diameter	Surface area	Volume	Density	Surface gravity	Escape velocity
	km	(Earth = 1)	(Earth = 1)	(kg m^{-3})	(Earth = 1)	(km s^{-1})
Earth	12,756	1.000	1.000	5.52 × 10^3	1.000	11.18
Moon	3,476	0.075	0.020	3.34 × 10^3	0.165	2.37

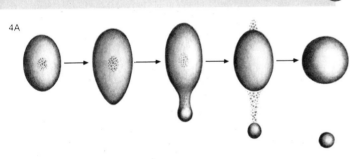

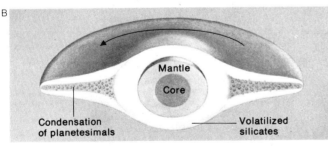

Mantle
Core
Condensation of planetesimals
Volatilized silicates

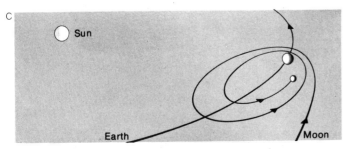

Sun
Earth
Moon

The Moon's Orbit

As observed from the Earth, the Moon appears to move in an elliptical orbit with the Earth at one focus, just as the planets move in ellipses around the Sun. In fact, it would be slightly more accurate to say that the Earth and Moon revolve together around the "barycenter", the center of gravity of the system, but since the Earth is much more massive than the Moon the barycenter lies within the terrestrial globe, 1,630 km below the surface. The eccentricity of the Moon's orbit is 0.0549, and its distance from the Earth (measured from center to center) ranges from 356,410 km at its closest approach or "perigee" to a maximum value of 406,679 km at "apogee"; the mean distance is 384,392 km. At apogee the apparent diameter of the Moon is only nine-tenths of that at perigee; the difference is inappreciable to the naked eye, but easy to measure. The line joining the perigee and the apogee is called the "line of apsides".

The plane of the lunar orbit is inclined to the plane of the ecliptic (defined by the Earth's orbit around the Sun) by an angle of 5°9'. The Moon therefore crosses the ecliptic at two points, once as it moves from south to north (the "ascending node") and once as it moves from north back to south (the "descending node").

However, the Moon's movement around the Earth is complicated in several ways by the gravitational pull of the Sun, which causes perturbations in the orbit. It is perhaps surprising to learn that the Sun's pull on the Moon is more than twice as strong as that of the Earth; viewed from outside the Solar System, the Moon would appear to revolve around the Sun, with an orbit that wobbles backwards and forwards about the Earth. (There is no chance, however, that the Earth and the Moon will part company, since the Sun attracts the two bodies almost equally.)

The perturbing influence of the Sun can be analyzed into six main effects. One is to cause periodic variations in the Moon's eccentricity, which oscillates between 0.044 and 0.067; this variation is called "evection". Similarly, the inclination of the orbit varies between 4°58' and 5°19'. Another effect is that the Moon's perigee advances in the same direction as the Earth's rotation, taking a period of 8.85 years to complete one revolution. A further consequence of the Sun's gravitation is that the line joining the Moon's nodes has retrograde motion along the ecliptic (in other words, in the opposite direction to the movement of the perigee); the period, known as the "nutation period", is 18.61 years. Finally, calculations of the position of the Moon must take into consideration the fact that the gravitational pull of the Sun on the Moon is weaker when the Moon is on the far side of the Earth than when it is situated between the Earth and the Sun (an effect known as "variation"), as well as allowing for the changes in the Earth's distance from the Sun during a year (the "annual equation").

The problem of giving a complete mathematical account of the Moon's orbit is one of the most difficult in astronomy. A more detailed description than has been given so far would have to take into account not only the effects of the Sun, but also such minor influences as the planets and the exact shapes of the Earth and the Moon; but for the purposes of giving a brief outline of the lunar orbit, these terms are sufficiently small to be ignored.

Phases of the Moon

Having no light of its own, the Moon shines only because of the light it reflects from the Sun: at any given moment the hemisphere turned away from the Sun will be dark. The apparent shape of the Moon in the sky (its "phase") therefore depends on what position it has reached in its orbit around the Earth. When it lies exactly between the Earth and the Sun, so that its sunlit hemisphere cannot be seen from Earth, the Moon is said to be "new"; shortly afterward a crescent will be seen. As the Moon moves further along its orbit, the size of the crescent increases as the Moon "waxes", and eventually half the surface turned towards the Earth is illuminated; this point is termed the "first quarter", the Moon having completed

1

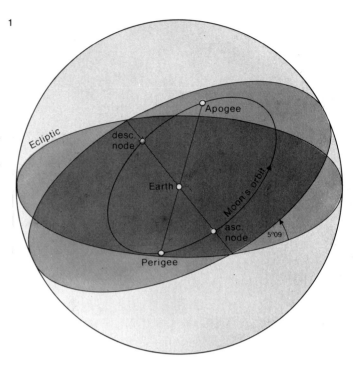

1. Orbit of the Moon
The Moon's orbit is shown projected against a celestial sphere centered on the Earth. The plane of the orbit is inclined to the plane of the ecliptic by an angle of 5° 9'. The Moon's nodes, defined by the two points at which the orbit crosses the ecliptic, move backward along the ecliptic at a rate of just over 19° per year. The line of apsides (joining the perigee and apogee) moves in the opposite direction with a period of 8.85 years.

2. Phases of the Moon
The phase of the Moon depends on the angle between the Sun, the Moon and the Earth. One cycle of lunar phases is completed in the interval between two successive similar alignments of the three bodies, as represented by the first and last positions shown here (**A**). The appearance of the Moon from Earth (**B**) is shown in relation to the corresponding position of the Moon in its path. The Moon takes slightly less time to repeat its initial position against the stellar background; in the penultimate position in diagram **A** (bold line) the Moon is shown one sidereal month after the start of the sequence.

2A

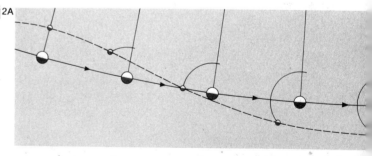

B

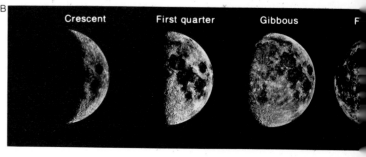

Crescent   First quarter   Gibbous   F

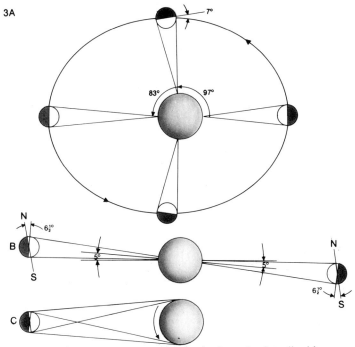

a quarter of its orbit. Next it becomes "gibbous", when more than half the sunlit hemisphere is visible, and eventually "full". The maximum magnitude of the full Moon is − 12.5 (*see* Glossary). Continuing its orbit, the Moon begins to "wane", once again becoming gibbous until it reaches the "third quarter", and then dwindling to a crescent. With the following new Moon, the cycle begins again. The "age" of the Moon is the number of days that has elapsed since the previous new Moon.

During the crescent phase, the unilluminated part of the Moon's disc is often clearly visible. This phenomenon is due to the reflection of light by the Earth, as was explained by Leonardo da Vinci.

The boundary between the bright and dark hemispheres of the Moon is known as the "terminator". Because the lunar surface is rough, the terminator is jagged in appearance, rather than smooth, and it often happens that a mountain, catching the first rays of the Sun over the horizon, appears as an isolated point of light slightly beyond the dark side of the terminator. The points or horns of the Moon when it is between new and full are known as the "cusps".

The Moon's period
It takes 27.32166 days for the Moon to complete one revolution, measured against the background of fixed stars. This period is known as a "sidereal month". There are, however, several other kinds of month. For example, because the Earth is itself moving around the Sun, the interval between one new Moon and the next is slightly longer than a sidereal month (*see* diagram 2) and is known as a "synodic month": it is equal to 29.53059 days. This period is also sometimes referred to as a "lunar month" or "lunation". There is also the "draconic" or "nodal" month (measured between successive passages through the same node), the "anomalistic month" (perigee to perigee) and the "tropical month" (successive conjunctions with the vernal equinox—*see* Glossary).

The Moon moves eastward against the stellar background by about 13° per day and therefore rises later on successive days; the time-difference between one moonrise and the next is known as the "retardation". Its average value is 50 min, but because the Sun and Moon do not move at uniform speeds, and because their orbits are inclined to the equator, this value varies throughout the year. In the northern hemisphere the retardation is greatest around March, when the angle of the Moon's path to the horizon is at its steepest; it is least around September, when the angle is shallowest. The full Moon that occurs nearest the time of the autumnal equinox (21 September) is called the "Harvest Moon" and rises shortly after sunset. The Sun is then on or near the equator, while the Moon is on the opposite side of the Earth crossing the equator from south to north. The subsequent full Moon, which also rises early in the evening, is called the "Hunter's Moon". In the southern hemisphere there is a six-month shift in the relevant dates.

Librations
The Moon's orbit around the Earth is "captured" or "synchonous": its rotational period is the same as its period of revolution and it therefore keeps the same hemisphere pointing towards the Earth all the time. However, because the lunar orbit is not circular, the velocity of the Moon varies, being greatest at perigee and least at apogee. At the same time, the rate of axial spin is constant, so that in the course of one lunar orbit it is possible to see a little more than half of the globe; this apparent "wobble" is termed the "libration in longitude". There is also a "libration in latitude" because of the inclination of the Moon's orbit, so that an Earth-based observer can see a short distance around alternate poles. Finally, there is a daily or "diurnal" libration, due to the axial rotation of the Earth itself: at moonrise an observer can see a little way beyond the eastern limb and at moonset a little beyond the western limb. The combined effect of these three librations is to allow slightly more than 59 percent of the Moon's surface to be examined from Earth.

3. Librations
The Moon's libration in longitude (**A**) is a consequence of the non-uniformity of its speed in its elliptical orbit around the Earth (exaggerated in this diagram). The Moon revolves at a constant rate, so that after one-quarter of the rotational period it has rotated through 90°; at the same time, however, it has moved through 97° of its orbit, so that from Earth the Moon's central meridian would appear to be displaced by about 7°. The libration in latitude (**B**) results from the combined effect of the 5° inclination of the Moon's orbit to the Earth's equator, and the $1\frac{1}{2}$° tilt between the Moon's axis and its orbit: as a result the Moon's axis of rotation is inclined by about $6\frac{1}{2}$° relative to the Earth's axis, so that the north and south polar regions are alternately tilted slightly towards the Earth. Finally, the diurnal libration (**C**) allows an Earth-based observer to view the Moon from a slightly different angle in the morning as compared with the evening, because his viewpoint changes with the Earth's daily rotation.

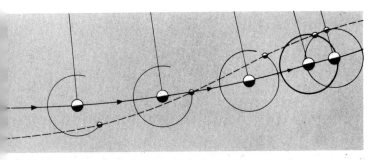

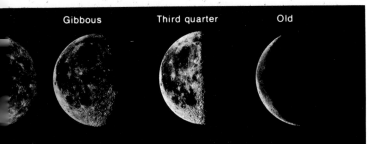

Gibbous   Third quarter   Old

Eclipses and Tides

It has been known since ancient times that the twice-daily rise and fall of the oceans was in some way related to the position of the Moon, but it was not until the problem was investigated by Sir Isaac Newton that an adequate explanation of the tides was given. Essentially, the tides occur as a result of the Moon's gravitational attraction, which causes slight bulges in the oceans, so that as the Earth rotates, every point on its surface is swept past the tidal bulges, alternately experiencing high and low tides. The reason for there being two tidal bulges, one on the side of the Earth facing the Moon and the other on the opposite side, is best explained by considering first a simplified model in which the Earth is completely covered with water and not rotating about its own axis (*see* diagram 4). The Moon's gravitational pull will be strongest on the closest hemisphere, causing the water to accelerate towards the Moon and heap up. Meanwhile, the water on the far side of the Earth will experience the weakest acceleration due to the Moon's gravity and will therefore be "left behind" as the Earth accelerates away from it, thus producing the second bulge. These two tidal bulges would move around the Earth with the same period as the Moon, causing high tides every 13.65 days. Since, however, the Earth itself rotates once every 24 hr, high tides occur at slightly shorter intervals, approximately every 12 hr 25 min.

In practice the situation is complicated by several other factors. Local conditions influence the exact time of the tides because of irregularities in the depth and form of the oceans. The Sun also influences the tides, either reinforcing or opposing the pull of the Moon. When the Moon is either full or new ("syzygy") its gravitational pull is combined with that of the Sun to produce particularly strong tides known as "spring tides"; when the Moon is in its first or third quarter ("quadrature") the pull of the Sun acts at right angles to that of the Moon so that the two partly cancel one another, giving weak or "neap" tides.

Moreover, frictional forces in the oceans produce an appreciable lag in the time taken by the waters to heap up, while the rotation of the Earth causes the bulge to be carried slightly ahead of the actual position of the Moon rather than directly beneath it. One important consequence of this frictional effect is that the Moon's gravity acts as a brake on the Earth's rotation, so that the Earth's period is gradually increasing. Studies of the growth lines in fossil corals suggest that 350 million years ago the Earth day was approximately 3 hr shorter than it is today, so that there were about 400 days in a year. Other studies, based on ancient eclipse records, indicate that the rate of increase of the Earth's period is about 0.0016 sec per century.

As the Earth's rate of spin decreases, energy is transferred to the Moon's orbit. The Moon is gradually accelerating, and as it does so its distance from the Earth increases. It may therefore be deduced that in the remote past the Earth and the Moon were closer together. (This is one of the reasons that first led G. H. Darwin to propose the "fission hypothesis"—*see* page 6.) In a similar way tidal forces acting on the slightly ellipsoidal shape of the Moon have slowed down the Moon's rotation over the ages, thus producing the present situation in which its period of axial rotation is exactly equal to its period of revolution. In fact, all major planetary satellites have similarly captured rotations.

Eclipses

Both the Earth and the Moon cast long conical shadows out into space. When the Earth passes through the Moon's shadow-cone, a "solar eclipse" occurs and can be seen on Earth by an observer situated in the area covered by the shadow; when in turn the Moon passes through the Earth's shadow there is a "lunar eclipse". Unlike an eclipse of the Sun, a lunar eclipse is visible from an entire hemisphere of the Earth.

During a lunar eclipse the Moon does not usually vanish completely, because some of the Sun's rays are bent or "refracted"

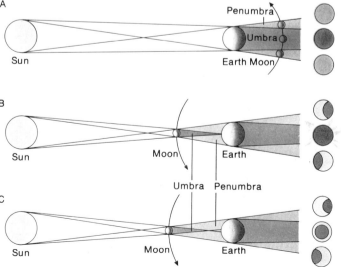

1. Lunar eclipse
This sequence of photographs shows the start of the total eclipse of the Moon that occurred on 24 June 1964; the Moon is shown moving into the shadow cone of the Earth. In 450 BC, Anaxagoras of Clazomenae reasoned that because the Earth's shadow on the Moon was curved, the Earth itself must therefore be spherical.

2. Conditions of an eclipse
The Moon's orbit is inclined to the plane of the ecliptic; an eclipse, either lunar or solar, can only occur if the points at which the Moon crosses the ecliptic lie on or near the line joining the Earth and the Sun (**A**). Otherwise the shadow cone of each body will pass above or below the surface of the other (**B**).

3. Types of eclipse
In a lunar eclipse (**A**), the Moon passes first into the "penumbra" of the Earth's shadow before reaching the darker "umbra". An eclipse of the Sun (**B**) only appears total from the limited region of the Earth's surface that is covered by the umbral shadow cone; from inside the penumbra the eclipse appears partial, and the solar corona is not seen. An annular eclipse (**C**) occurs when the Moon is near apogee, and its shadow cone does not reach the Earth's surface.

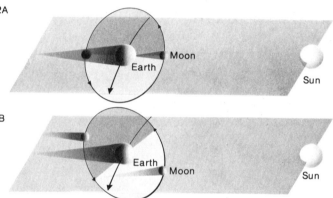

onto the lunar surface by the atmosphere surrounding the Earth. The Moon's appearance during an eclipse, therefore, depends on the conditions in the Earth's atmosphere, which can give the Moon a reddish or coppery color; on other occasions the Moon may be so dark as to be invisible to the naked eye.

An eclipse need not be "total"; if the disc of the Sun or Moon is not entirely obscured by shadow because the bodies do not line up exactly, then the eclipse is said to be "partial". Thus a partial eclipse of the Sun would be seen by an observer standing on Earth in the area covered by the Moon's "penumbra", the region of partial shadow that surrounds the main cone of shadow or "umbra" (*see* diagram 3). In the case of solar eclipses, there is a third type of eclipse, called "annular", in which the Moon is directly in line with the Sun, but its disc appears too small to produce a total eclipse, so that the Sun's rim can be seen forming a bright ring around the Moon. Annular eclipses occur because the distance of the Moon from the Earth is sometimes greater than the length of the Moon's umbra, so that the conical shadow fails to reach any point on the Earth's surface.

A solar eclipse can only occur when the Moon is new, a lunar eclipse only when it is full. However, because the Moon's orbit is inclined to the plane of the ecliptic, eclipses do not occur at every full or new Moon. In order for there to be an eclipse, the Moon must be at or near one of its nodes. If the angle between the line of the nodes and the Sun or the Moon is greater than $18°\ 31'$ there can be no solar eclipse, while if this angle is greater than $12°\ 15'$ there can be no total lunar eclipse. The maximum number of eclipses of both kinds that can occur in a year is seven, the minimum, two. Eclipses tend to come in twos or threes, a lunar eclipse always being preceded or followed by a solar eclipse. The duration of eclipses is variable, depending on the exact geometry of the Earth, the Sun and the Moon. For a lunar eclipse totality can last for as long as 1 hr 44 min; for a solar eclipse the maximum is 7 min 40 sec. Annular eclipses can last longer.

The position of the Moon's nodes in relation to the Sun is clearly of particular importance in determining whether or not an eclipse will occur. Because of the retrograde motion of the nodes (*see* page 8), the ascending node will come into conjunction with the Sun in slightly less time than it takes the Sun to complete one orbit: in other words, the synodic period of the ascending node is less than a year. This period, known as the "eclipse year", is equal to 346.62003 days. The Sun will be in line with one of the nodes twice during this interval.

The Saros

It was discovered by the ancient Babylonians that eclipses with very similar circumstances tend to recur after an interval of about 6,585 days, equal to 18 years 10 or 11 days (depending on the number of leap years). This period is called the "Saros" and can be used as a reasonable guide to predicting eclipses. The Greek philosopher Thales of Miletus is reported to have used the Saros to predict the eclipse of 28 May 585 BC.

The method works because of a mathematical coincidence. Nineteen eclipse years (6,585.78 days) happens to be almost exactly equal both to 223 synodic months (6,585.32 days) and to 239 anomalistic months (6,585.54 days). In other words, the same configuration of the Sun, the Moon and the Earth is almost exactly repeated after one Saros interval. For example, on 29 January 1953 there was a total lunar eclipse, visible from England. One Saros later, on 10 February 1971, there was another total eclipse, although it was not fully visible from England because the Moon set before the eclipse was over.

Because the three periods are not exactly equal, each eclipse in a given series differs slightly from the previous one, and eventually the cycle comes to an end. At any given moment several Saros series will be in progress, overlapping one another.

4. Cause of the tides
The phenomenon of the tides may be simplified by ignoring frictional forces and imagining a non-rotating Earth that is covered with water. Under such circumstances the water on the side of the Earth closest to the Moon would experience the strongest gravitational attraction, while those on the far side would be weakly attracted, thus two tidal bulges would be formed (**A**). These would follow the Moon as it orbited the Earth. Since, however, the Earth is rotating (**B**), its surface is swept past the bulges twice daily. At the same time frictional forces cause the bulges to be carried round slightly ahead of the Moon, rather than directly below it.

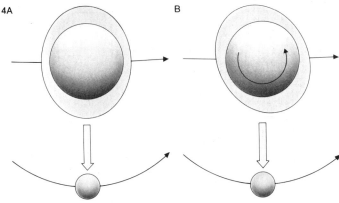

4A   B

5. Spring and neap tides
The Sun as well as the Moon influences the Earth's tides, although to a lesser extent. As a result the height of the tides varies with the lunar cycle: at new or full Moon (**A**) the effect of the Sun reinforces that of the Moon, producing strong "spring" tides; at the first and third quarters (**B**) the Sun's attraction partially cancels that of the Moon, giving weak "neap" tides. The height of the water level at a given location fluctuates as shown (**C**) during half a lunation.

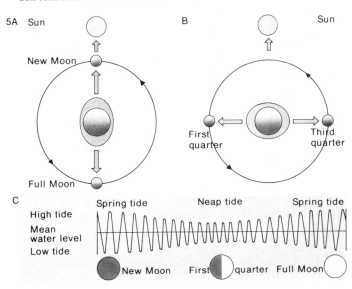

5A   Sun   B   Sun

New Moon

Full Moon

First quarter   Third quarter

C

Spring tide   Neap tide   Spring tide

High tide
Mean water level
Low tide

New Moon   First quarter   Full Moon

History of Lunar Observations

The main dark areas of the Moon are easily visible with the naked eye, and the first of all lunar maps seems to have been that of the British scientist W. Gilbert in or about 1600; it was published posthumously in 1651, and some of the details are plainly recognizable. The first telescopic map was the work of Thomas Harriott, one-time tutor to Sir Walter Raleigh, in July 1609, but the earliest serious work was carried out from 1610 by Galileo, who made drawings of special features and also attempted to estimate the heights of the lunar mountains by measuring the lengths of their shadows. His results gave altitudes which were rather too high, but they were at least of the right order, and showed the Moon's mountains are higher, comparatively, than those of the Earth.

In Wales Sir William Lower is said to have used a telescope to look at the Moon around 1611. His drawings have not been preserved, but he made the picturesque comment that the full moon looked rather like "a tart which his cook had made".

Nomenclature

The pioneer attempts to draw up a useful nomenclature were made by Langrenus in 1645 and by Hevelius in 1647. Hevelius, a city councillor of Danzig (now Gdansk), used geographical analogies; for instance, the crater now called Plato was called "the Greater Black Lake". His nomenclature never became popular, and it is said that after his death the copper engraving of his lunar map was melted down and made into a teapot.

Hevelius's nomenclature was superseded by that of the Italian Jesuit Giovanni Riccioli, who published a map in 1651 based on observations by his pupil, Grimaldi. Riccioli named the mountains after Earth ranges such as the Alps and Apennines, and the craters after famous persons, usually (though not always) astronomers. His system has survived and has since been extended to cover minor features and also those on the Moon's far side. It is reasonably satisfactory, and will certainly never be altered now, but inevitably later scientists have had to be allotted less prominent features. Newton's name, for example, is attached to a formation not far from the south pole. Moreover, Riccioli was not impartial in his choices: he did not believe in the Copernican System (according to which the Earth moves around the Sun) and therefore "flung Copernicus into the Sea of Storms"; Galileo was similarly allocated a very small insignificant crater. Grimaldi and Riccioli himself, meanwhile, have large, prominent walled plains near the western limb, identifiable under any conditions of solar illumination because of their dark floors. The maria were given "romantic" names like "Serenitatis" and "Imbrium" to distinguish them from the craters. At a much later date, two small maria near the limb were named after astronomers (Mare Humboldtianum and Mare Smythii). There are also some unexpected names; Julius Caesar has a crater of his own, presumably because of his association with calendar reform. One crater is named Hell. This does not, however, indicate any exceptional depth but honours a Hungarian astronomer, Father Maximilian Hell.

Development of selenography

In 1775 the German astronomer Tobias Mayer produced a small but fairly accurate lunar map and also introduced a system of coordinates; but the real "father of selenography" was Johann Hieronymus Schröter, chief magistrate of the little town of Lilienthal, near Bremen. Schröter built a private observatory and studied the Moon and planets from 1778 until his observatory was destroyed by the invading French soldiers in 1813. He made hundreds of drawings of lunar formations and was the first to make detailed observations of the crack-like features known as clefts or rills. Despite his admittedly clumsy draughtsmanship his work was of the utmost value, and his name is commemorated in "Schröter's Valley". The next real progress was due to a Berlin banker, Wilhelm Beer, and his colleague, Johann Mädler. Using the small refractor

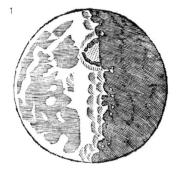

1. Galileo's observations, 1609
Ptolemaeus (lower center), Copernicus (white patch, left center) and Mare Imbrium (upper left) are recognizable in these drawings from *Sidereus Nuncius*.

2. Lunar map by Hevelius, 1647
Hevelius produced a map showing many recognizable features, but his system of nomenclature, based on geographical analogies, has not survived.

3. Lunar map by Riccioli, 1651
Riccioli's map (*Almagestum Novum*) was better than that of Hevelius. Riccioli introduced the system of naming features after personalities. Many of the features shown on the map are recognizable; even the ray-systems from the craters Tycho, Copernicus and Kepler are shown. Riccioli also drew the Moon at different phases.

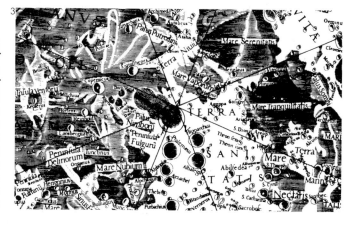

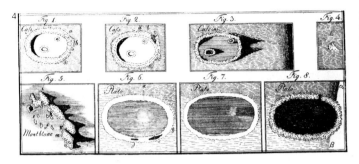

4. Lunar details by Schröter, 1791
Schröter never produced a complete map, but he made thousands of drawings. These sketches show Mont Blanc and the craters Cassini and Plato.

5. Map by Beer and Mädler, 1837
This map was a masterpiece of careful observation. It remained the standard work for over 40 years after being compiled, and in general is remarkably accurate.

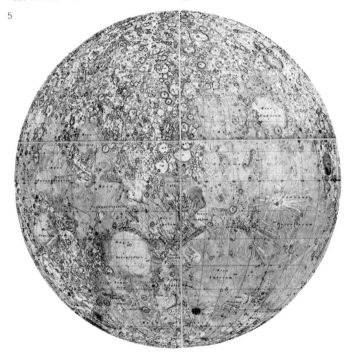

6. Detail by J. Schmidt, 1878
Schmidt's map, based on preliminary work by Lohrmann much earlier, was extremely detailed and superseded that of Beer and Mädler.

7. Far side preliminary chart, 1959
This early chart was drawn from the photographs obtained from Luna 3. Some features are recognizable, such as Tsiolkovsky; others were misinterpreted.

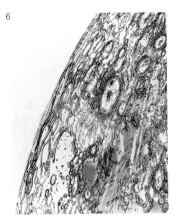

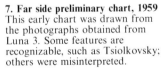

in Beer's observatory, they produced in 1837 a lunar map which was a masterpiece of careful, accurate observation, and followed it up with a book, *Der Mond*, containing a detailed description of every named feature. Unlike Schröter, they correctly believed the Moon to be dead and changeless. Interest was greatly stimulated in 1866, when Julius Schmidt, a German who had emigrated to Greece and become Director of the Athens Observatory, announced that the crater Linné, recorded by Beer and Mädler on the grey plain of the Mare Serenitatis, had disappeared to be replaced by a white patch. Even though it now seems unlikely that any real change occurred in Linné, the repercussions of Schmidt's announcement were far-reaching and the report was probably directly responsible for a number of new maps. Schmidt himself published an elaborate version in 1878, based on a chart begun much earlier by Wilhelm Lohrmann, a Dresden land surveyor; in England Edmund Neison produced a map and a book, and the British Astronomical Association, formed in 1890, planned an even larger chart, although little of it was ever actually completed.

Modern selenography
The first photograph of the Moon was taken in 1840 by J. W. Draper, and from the 1870s photography became an essential basis of all lunar mapping. Photographic atlases were produced—notably by Loewy and Puiseux, at the Paris Observatory, in the 1890s and by W. H. Pickering in 1904. Pickering, one of the few professional astronomers to take an interest in the Moon at that time, showed each part of the Moon under five different conditions of illumination, though it is true that the pictures are of poor quality by modern standards. The members of the Lunar Section of the British Astronomical Association were both enthusiastic and skilful, and the outline chart drawn by its first Director, Thomas Gwyn Elger, is still useful. It was compiled mainly from direct visual observations, but with a photographic basis.

In 1840, when photography was just beginning, the French astronomer François Arago stated that by this new technique it would be possible to complete lunar mapping in a matter of weeks. He was wrong; it took more than a century and a quarter, but today the Moon is better mapped than some regions of the Earth.

Before the Space Age, study of the Moon was regarded as essentially an amateur province. Further maps were produced, notably by Walter Goodacre in 1930 and by H. P. Wilkins in 1946, and were probably better than the official chart published in 1935 by the International Astronomical Union, based on photographic micrometrical measurements by W. H. Wesley and M. A. Blagg. As flight to the Moon became a real possibility, however, professional observatories took over the task of lunar cartography, although they owed much to the older amateur work.

Particular attention had always been paid to the libration regions of the Moon (*see* page 9), which are highly foreshortened and therefore difficult to chart. It was possible for amateurs to make interesting discoveries: for instance, the great ringed structure known as the Mare Orientale was first seen (and named) by Wilkins and the present writer just after the war.

During the 1950s elaborate photographic atlases were compiled, mainly by American astronomers at the Lunar and Planetary Laboratory at Tucson, Arizona, and these naturally made the older charts virtually obsolete. The system of nomenclature was refined, and new names were added where appropriate. By 1959, when the first unmanned lunar probes were launched, maps of the Moon's Earth-turned hemisphere were reasonably good except in the libration regions, but obviously they could not show the very small features. The final phase began with the American Orbiters of 1966–68. These were put into closed orbits around the Moon and sent back thousands of highly detailed photographs, covering almost the whole of the surface. Without them, it is not likely that the manned Apollo missions could have been attempted.

Unmanned Lunar Probes

As long ago as the second century AD, the Greek satirist Lucian of Samosata wrote a story, the *True History* (described by the author himself as being made up of nothing but lies from beginning to end), in which the intrepid astronauts were sailors whose ship was caught up in a waterspout and hurled onto the Moon. Much later came stories such as the *Somnium* or "Dream", written by the great astronomer Johannes Kepler, where the journey was accomplished with the aid of friendly demons. But it was only with the development of rockets as serious research tools that lunar flight became a real possibility.

Russian lunar missions

The age of space exploration began on 4 October 1957 with the launching of the Russian satellite Sputnik 1. Less than two years later, on 2 January 1959, the Soviet researchers dispatched their first Moon probe, Luna (or Lunik) 1, which passed within 5,955 km of the Moon. No pictures were obtained, but some useful information was sent back, notably confirmation that the Moon has no detectable overall magnetic field. Next, on 12 September of the same year, came Luna 2. This was a "hard-lander": it crashed down, destroying itself and without sending back any data, but it did at least show that the Moon was within range. The real triumph came with Luna 3, launched on 4 October 1959—exactly two years after Sputnik 1. It went right around the Moon, at a minimum distance of 6,200 km, taking the first photographs of the far side. When the spacecraft came back to the neighborhood of the Earth, on 18 October, the pictures were scanned by a miniature television camera and transmitted to Earth, where they were picked up by the waiting Russians. They were released to the world on 24 October, giving the first direct views of the mysterious regions on the far side of the Moon.

By today's standards the pictures were blurred and deficient in detail, but they showed that the far side is essentially similar to the highlands on the familiar hemisphere, although deficient in the dark plains or maria (*see* pages 26–27). There were some identifiable features, notably the so-called Mare Moscoviense or "Sea of Moscow", and the vast, dark-floored walled enclosure which was named in honor of Konstantin Tsiolkovsky, the great Russian rocket pioneer. Ray craters were also seen. Inevitably there were some misinterpretations: for example, a long chain of peaks christened the "Soviet Mountains" does not exist at all. Whether Luna 3 was intended to send back further data is not known. Contact with it was abruptly lost and never regained.

The next four Lunas were unsuccessful, but in July 1965 Zond 3 obtained further pictures of the Moon's far side, from a minimum distance of 9,219 km. Luna 9, launched on 31 January 1966, made a successful soft landing on the grey plain of the Oceanus Procellarum and sent back pictures direct from the lunar surface, finally disposing of the notion that the lunar maria might be covered with deep layers of very soft dust incapable of supporting the weight of a spacecraft.

The next three Lunas (numbers 10, 11 and 12) were put into closed paths and sent back useful data as well as pictures; Luna 13, sent up on 21 December 1966, was another successful soft-lander, capable of carrying out soil analyses and other valuable researches. Since then there have been more than a dozen further unmanned vehicles, listed in Table 1, a few of which have been of special note. For instance, Zonds 5, 6 and 7 went right around the Moon and then were brought safely home to Earth. Lunas 16 and 20 landed, obtained samples of lunar material, and brought them back for analysis. The special value of these latter probes was that they demonstrated the feasibility of remote sampling even to the extent of digging core samples some way into the surface. These techniques will be vital to planetary exploration, where manned flights are not yet possible. It was also useful that the probes came down in areas untouched by the Americans—in the Mare Fecunditatis and

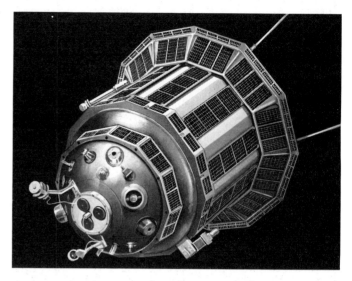

Luna 3 (USSR): first flight around the Moon, 1959

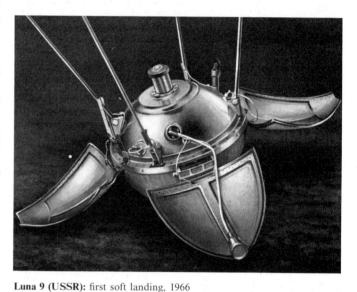

Luna 9 (USSR): first soft landing, 1966

Lunokhod 1 (Luna 16—USSR): crawled on the Moon, 1970

Ranger 7 (USA): crash landing, 1964

Lunar Orbiter 5 (USA): photographic coverage, 1966

Surveyor 3 (USA): soft landing, 1967

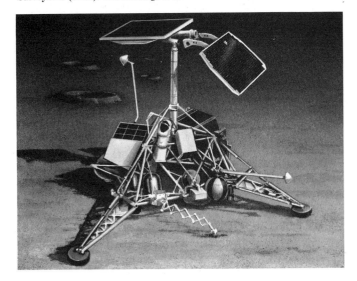

south of the Mare Crisium respectively—and provided information about unexplored parts of the Moon.

Luna 16 was a new departure. It carried a "crawler", Lunokhod 1, which looked quite extraordinary, but which proved to be a true triumph. Remote controlled from Earth, it crawled around the Moon, covering a total distance of 10.5 km, sending back 20,000 pictures and covering photographically an area of 80,000 m²; it operated for 11 months after its arrival on 11 November 1970 in the general area of the Sinus Iridum. It was followed by Lunokhod 2, carried by Luna 20, which was equally successful and obtained data from the Le Monnier region. The present series came to a temporary halt with Luna 24, which was launched on 9 August 1976, landed in the Mare Crisium on the 18th, drilled into the surface to collect a 2 m core sample before returning to Earth, arriving back on 22 August.

American lunar missions
It is significant that the Russians have concentrated exclusively on automatic vehicles, which adds force to the argument that there never was a "space race" and that the Soviet and United States programs were planned along different lines. At first the Americans were confronted with a series of failures. Either their probes failed after take-off, or missed the Moon altogether, or else they went out of contact, or crash-landed out of control. The first real success was with Ranger 7, which hit the Moon on 31 July 1964. It was not designed to survive impact, but before landing it sent back 4,316 photographs of a region in the Mare Nubium which has since been named the Mare Cognitum or "Known Sea". These pictures were very detailed and far better than those from the Soviet vehicles; so too were the pictures from Ranger 8 (20 February 1965, in the Mare Tranquillitatis) and Ranger 9 (24 March 1965, inside the great walled plain Alphonsus).

The Ranger project was followed by two further series of unmanned probes, the Lunar Orbiters and the Surveyors. The Orbiter program consisted of a total of five probes, all of which were launched within a year, starting on 10 August 1966. One of their main objectives was to help in the selection of suitable landing sites for future manned missions, with particular attention to the equatorial regions.

All worked faultlessly, and even today many of the thousands of photographs sent back await detailed examination. (Many of the photographs in this atlas came from the Orbiter missions.) There were other invaluable results. For instance, slight irregularities in the movements of Orbiter 5 led to the discovery of the so-called "mascons" (*mass con*centrations), regions of higher density than average below the lunar crust (*see* page 20). They also studied micrometeoroids and radiation effects. After completing their programs, the Orbiter spacecraft were deliberately crashed onto the Moon so as to avoid radio interference with later missions. Orbiter 5, the last of the series, was given a final command to crash-land on 31 January 1968, bringing this particular phase of lunar exploration to a conclusion.

Up to 1966 the theory of deep dust-drifts was still taken seriously in the United States and there was considerable relief when the soft-landing of Luna 9 showed it to be wrong. By then the American Surveyor program, also designed for soft landing, was in the last stages of planning, and on 30 May 1966 the first vehicle in the series went on its way. It came down north of the crater Flamsteed and returned over 11,000 splendid photographs, which showed the Moon's surface in much greater detail than ever before. The Surveyor series was continued until January 1968; Surveyors 1, 3, 5, 6 and 7 were successful, while numbers 2 and 4 were not. The pictures obtained were quite remarkable in their detail and high resolution. Surveyor 7, the last of the series, came down on the northern rim of the ray crater Tycho, returning 21,000 photographs, as well as a vast amount of scientific data.

15

Manned Flights

The Apollo program was initiated in the early 1960s, with the enthusiastic support of President John F. Kennedy. The procedure was modified many times in the planning stages, but it was finally decided to carry out the landing in a Lunar Module (LM), taken into lunar orbit by the main spacecraft; of the three members of the crew, one would remain in orbit while the other two made the actual landing. The first manned circumlunar flight was that of Apollo 8 in December 1968. The Lunar Module was then tested in Earth orbit (Apollo 9), and in May 1969 there was another flight around the Moon (Apollo 10), so that the LM could be tested as thoroughly as possible without an actual touch-down.

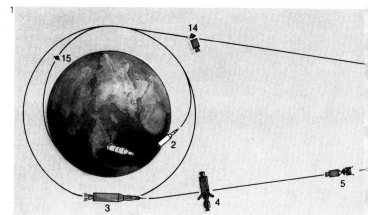

The culmination of all the preparatory tests was Apollo 11, which was launched on 16 July 1969 from Cape Kennedy. (Cape Kennedy has now reverted to its original name of Cape Canaveral.) The crew consisted of Neil Armstrong, Edwin Aldrin and Michael Collins; Collins was to remain in the Command Module while Armstrong and Aldrin made the descent to the lunar surface. The actual landing was carried out manually to enable the astronauts to avoid boulders. The Lunar Module touched down in the Mare Tranquillitatis at 0°7N, 23°E; first Armstrong, then Aldrin, set foot on the bleak landscape of the Mare Tranquillitatis early on 21 July; the two astronauts remained outside the LM for over 2 hr, setting up what is known as the ALSEP or Apollo Lunar Surface Experimental Package.

The scene was very much as had been expected, and Aldrin's description of it as "magnificent desolation" could hardly have been bettered. On emerging from the Module, the first task was to collect samples. Armstrong reported that "the surface is fine and powdery . . . I can pick it up loosely with my toe. It does adhere in fine layers like powdered charcoal to the sole and sides of my boots. I can only go in a small fraction of an inch."

Of the experiments in the ALSEP, the first to be set up was the equipment designed to collect particles from the "solar wind", a continuous stream of low-energy particles emitted by the Sun. There was a "laser mirror", composed of special crystals to reflect laser beams sent out from Earth, thus providing a means of measuring the exact distance of the Moon; and there was a seismometer, to measure any "moonquakes" or tremors in the lunar crust. The astronauts also collected 21.75 kg of rock samples to take back to Earth for analysis. In addition to the scientific equipment the astronauts set up a U.S. flag with a special support to hold out the flag in the absence of any atmosphere. About $21\frac{1}{2}$ hr after they had landed they began the ascent to join the Command Module. The take-off went smoothly, and in less than 4 hr Armstrong and Aldrin were back with Collins in the Command Module. The Lunar Module was jettisoned and the return journey to Earth began. Finally, on 24 July the astronauts splashed down safely in the Pacific Ocean. The eventual landing was almost exactly in the prearranged position.

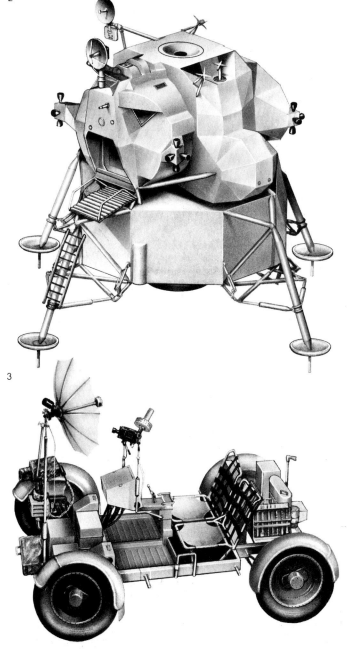

At that stage the astronauts and returned samples were strictly quarantined. Although the chances of bringing back any harmful materials from the Moon were negligible, the precaution was wise. Only when it had been established beyond all doubt that the Moon was sterile was the quarantine procedure relaxed. If men ever visit Mars (which, unlike the Moon, has an appreciable atmosphere) even more strict precautions will have to be taken, perhaps including quarantine in an orbiting space laboratory, although the preliminary evidence strongly suggests that there is no "life" on Mars either.

Later Apollo missions

Less than four months after the triumph of Apollo 11, a second manned mission was on its way. Apollo 12 touched down in the Oceanus Procellarum on 19 November 1969; it was a precision landing close to the old unmanned probe Surveyor 3 and the astronauts, Charles Conrad and Alan Bean, actually secured pieces of Surveyor and brought them home for study. They remained

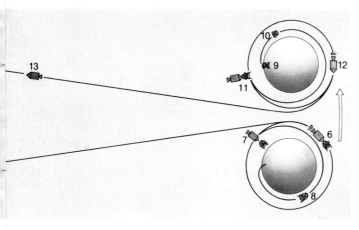

1. Flight plan of Apollo 11
(1) Lift-off from Cape Kennedy.
(2) Separation of first stage.
(3) Engine ignition to take rocket out of Earth-orbit and begin journey to the Moon. (4) Separation of Command and Service Module from Lunar Module. (5) Engine ignition of Service Module. (6) Lunar orbit insertion. (7) Separation of Lunar Module from Command Service Module. (8) Lunar Module descent engine ignition. (9) Start of return journey; Lunar Module ascent. (10) Maneuver to prepare for rendezvous with Command Service Module. (11) Docking. (12) Transfer of crew and equipment from Lunar Module to Command Service Module. (13) Return journey. (14) Service Module jettisoned. (15) Re-entry; communication blackout. (16) Final touchdown in the Pacific Ocean. The entire journey lasted 8 days, including the $21\frac{1}{2}$ hours spent on the lunar surface.

2. Lunar Module (LM)
All the Apollo Lunar Modules were of the same type. The lower stage served as a launch pad when the astronauts took off from the surface of the Moon.

3. Lunar Roving Vehicle (LRV)
LRVs, carried on the last three Apollo missions, could travel at about 15 km hr^{-1} on level ground.

4. Apollo Lunar Surface Experimental Package (ALSEP)
The general layout of an ALSEP is shown here; the equipment is not all drawn to the same scale. (1) ALSEP central station; coordinates the experiments and relays their signals back to Earth. The rod-like feature is an antenna. (2) Radioisotope Thermal Generator (RTG). Approximately 70 W of power for the experiments was provided by radioactive decay. (3) Passive Seismometer, covered by an insulating blanket, monitored small vibrations of the Moon's surface. (4) Lunar Surface Magnetometer measured the magnetic field along three axes. (5) Solar Wind Spectrometer, measuring the energy, density, direction and variations with time of the Solar Wind. (6) Heat Flow Experiment, with two probes placed in holes drilled in the lunar surface. (7) Suprathermal Ion Detector and Cold Cathode Ion Gauge; two experiments studying the lunar atmosphere.

outside the LM for longer than the Apollo 11 crew (a total of 7.6 hr), covering 1.4 km, and the amount of time they spent on the Moon was also greater ($31\frac{1}{2}$ hr). Conrad and Bean set up an improved ALSEP powered by nuclear energy. After departure the jettisoned Lunar Module was deliberately crashed onto the Moon, setting up vibrations which continued for almost an hour and making a crater 6 m wide.

Apollo 13, of 11 April 1970, was aimed at the Fra Mauro region in the Mare Nubium, but was nearly a disaster. There were complications from the start. At a late stage it was found that all three astronauts (James Lovell, Fred Haise and Thomas Mattingly) had been in contact with German measles. Lovell and Haise were found to be immune but Mattingly was not, and the decision was taken to replace him as the pilot of the Command Module by Jack Swigert. There was a slight loss of power during the initial ascent, and then, on 13 April, there was an explosion which made a gaping hole in the side of the vehicle and put the main motors permanently out of action. The lunar landing was abandoned, and it was only by inspired improvisation that tragedy was averted. The power of the Lunar Module was used to swing the vehicle around the Moon and bring it home; had the explosion happened on the return journey, after the Lunar Module had been jettisoned, there would have been no hope at all. It was a salutary reminder that space is a dangerous environment.

There followed four further successful missions. Apollo 14 came down near Fra Mauro, and the astronauts took a "lunar cart" to help in collecting samples. Apollo 15 was brought down in the Hadley–Apennines area and was even more ambitious, as the astronauts (Scott and Irwin) had a special Lunar Roving Vehicle which enabled them to drive around, covering almost 30 km and taking them close to the great chasm known as Hadley Rill. By this time the purely scientific experiments were becoming more extensive and more elaborate, and in addition a sub-satellite was sent out from the orbiting section of the Apollo to undertake particle and field measurements. Apollo 16 took astronauts Young and Duke to the highlands of Descartes; this was a new type of landscape, and the results were of exceptional importance, despite the failure of a device intended to measure the heat outflow of the Moon's interior.

With Apollo 17, of December 1972, there was a new development. All previous Moon-travellers had been astronauts who had been trained in science specifically for the purpose, but for this last mission the process was reversed; Dr. Harrison Schmitt, a professional geologist, was trained as an astronaut and made the journey. His specialized knowledge was extremely useful, and there were some surprises—particularly the discovery of "orange soil", at first thought to be indicative of comparatively recent volcanism, but later found to be due to the very ancient colored glassy particles. The landing was made in the Taurus–Littrow area, a valley between high massifs on the borders of the Mare Serenitatis.

Importance of manned flight
It has sometimes been claimed that the sending of astronauts to the Moon was unnecessary, and that all the information required could have been collected by unmanned probes. Yet there is little or nothing to be said for this view. Unmanned vehicles have been essential, but if a base is to be set up on the Moon, as is an ultimate aim, the Apollo missions were of paramount importance. They have shown the way, and they have extended knowledge beyond all recognition. They have also confirmed that walking about on the lunar surface is by no means difficult. Moreover, it must be remembered that it was not merely a question of going to the Moon, exploring and coming home. Various experiments were set up, and information continued to be sent back for almost five years after the departure of the last astronauts. The legacy of Apollo is the 380 kg collection of samples, which is still providing information about the Moon.

Features of the Moon

In his very early observations of the Moon, dating from the winter of 1609–1610, Galileo recognized two essentially different types of lunar terrain. On the one hand, there are large grey plains, and on the other, heavily cratered highlands which are distinctly lighter in appearance. Originally the plains were thought to be seas and accordingly were named "maria", while the highlands became known as "terrae" or continents. It was soon realized, however, that there could be no open water on the Moon, but the names have nevertheless been retained, and as recently as the mid-1960s some authorities still believed that the maria were old oceans which had dried up and been covered over.

Other theories have also been proposed and rejected. One curious idea was that the entire surface was covered with a thick layer of ice. However, in 1890 the fourth Earl of Rosse was able to measure the tiny quantity of heat received from the Moon and showed that the daytime temperature is very high. Measurements from the Apollo 17 site indicate that it can reach 384 K, although admittedly the nights are very cold (below 103 K). According to another theory, proposed by T. Gold in 1955, the maria were covered with dust layers kilometers deep, so that if a spacecraft were to attempt to land, it "would simply sink into the dust with all its gear"; the dust would flow downhill, accumulating in the lowest-lying areas. This theory gained a considerable degree of respectability until the successful soft landing of Luna 9 in 1966, when it was finally discarded.

Surface features

The entire surface of the Moon is dominated by walled circular enclosures known as "craters", although in many cases "walled plains" would be a better term. Some of them are truly immense. The great enclosure Bailly, for example, close to the Moon's south-western limb as seen from the Earth (and therefore very fore-shortened), is almost 300 km in diameter. Craters of more than 100 km across are common, and many of them have either central peaks or groups of central elevations. Smaller craters are remarkably numerous; they cluster thickly in bright highlands and are also to be found on the maria. Their basic form is circular, but they frequently break into each other and in many cases are distorted out of all recognition, while on the maria are found barely recognizable "ghost craters" such as Stadius, with ramparts rising to only a few meters or tens of meters above the outer surface. Some craters, notably Tycho and Copernicus, are the focal points of systems of "bright rays", best seen under conditions of high solar illumination, and which are at their most conspicuous near the time of the full Moon. "Plateaus" are much less common, but there is one splendid example, Wargentin, 89 km in diameter, adjoining the vast walled plain Schickard in the south-west part of the Moon. This crater is filled with lava almost to its brim.

Isolated peaks are common—Pico and Piton, on the Mare Imbrium, are excellent examples—and there are high chains of mountains, although in general these chains form the borders of the regular maria (for instance, the Apennines and the Alps make up part of the border of the huge Mare Imbrium). Some of these peaks are extremely lofty and may exceed the heights of terrestrial ranges such as the Rockies; since the Moon is a much smaller world than the Earth, the relative heights of the lunar peaks are much greater. Originally their heights were measured by their shadows, but today more refined methods are available. Since there is no sea-level on the Moon, the altitudes have to be related to the mean radius of the globe.

Valleys are of various kinds. The celebrated Alpine Valley slices through the mountains, while the so-called Rheita Valley, in the south-eastern highlands, is made up of craters which have run together. Then there are crack-like features or "rills", known also as rilles or "rimae". Some of them, such as those associated with Hyginus and Ariadaeus, are visible with very small telescopes under

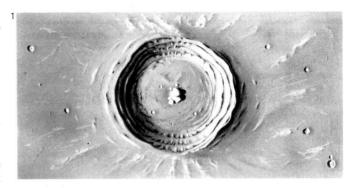

1. Terraced crater
The inner walls of this type of crater show well-defined terraces. The floor is sunken and there is a prominent central peak. Such craters are often surrounded by ejected material. Examples: Theophilus and Alpetragius.

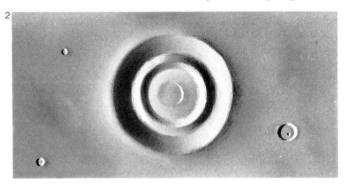

2. Concentric crater
Craters with multiple rings are found in all sizes on the Moon. The nested form of the smaller examples of such craters has been compared to nested volcanoes on Earth. Examples include Taruntius, Hesiodus, Marth.

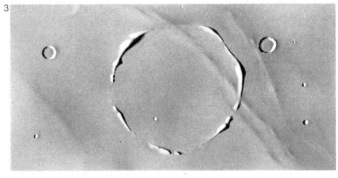

3. Ghost crater
Some craters appear to have been flooded by mare material so that their walls are barely traceable.

4. Ray-crater
Certain craters are distinguished by systems of bright rays which may extend for great distances.

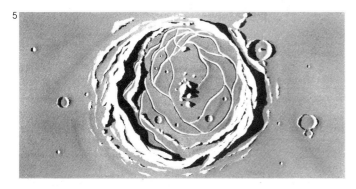

5. Rills
On the floors of some craters there are very extensive complicated systems of narrow furrows called "rills" or "rimae". Examples of such rills are found in Gassendi, Hevelius and Wilhelm Humboldt.

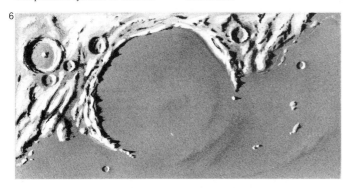

6. Lunar bay
Bay-like features sometimes occur where a crater bordering on a mare has been partially flooded so that part of its wall is destroyed. Examples include Fracastorius, Hippalus and Lemonnier.

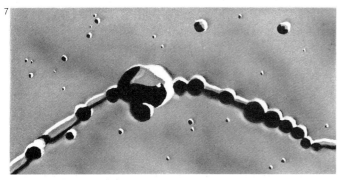

7. Crater-chain
This illustration shows one of the varieties of lunar rill; the so-called Hyginus Rill is a good example.

8. Isolated peak
Mountain peaks sometimes take the form of regular dome-like shapes, but many are irregular.

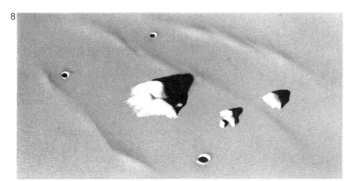

suitable conditions of illumination, and in many cases there are complicated systems, such as those of Triesnecker, in the Mare Vaporum–Sinus Medii area, and Hippalus, on the border of the small comparatively regular Mare Humorum. However, not all rills are regular and crack-like. Many—notably those of Hyginus—are composed of chains of small craterlets that have run together, often with the destruction of their common walls; they have been likened to "strings of beads". Finally there are "faults", the most striking example of which is the Straight Wall in the Mare Nubium (which, incidentally, is not straight, and is not a wall). The surface drops sharply to the west, so that the feature appears as a dark line before full Moon (on account of its shadow) and as a bright line afterward (because its steep face is being illuminated by the Sun).

Grid system
The Moon is criss-crossed with faults and ridges that form an important pattern known as the "grid system". Sometimes the faults are due to almost vertical slippages, while in other cases the movements have been more nearly horizontal. In many cases the ridges form parts of the walls of craters, though often enough the craters themselves are so broken and distorted that they are hard to identify. It has been pointed out by various authorities that there are two principal "families" of faults and ridges, running in specific directions and almost at right angles to each other. This makes up the main grid system, although it is complicated by minor families such as those of the Alps and Caucasus regions. The presence of the grid system indicates that the Moon's crust was subjected to strain over a long period, and it has also been found that small craters in some regions are non-circular, with their longest axes lying in the direction of one or the other of the families of the main grid.

Far side of the Moon
The far side, totally unknown before the flight of Luna 3 in 1959, is distinctly different from the familiar near side. The most striking difference is the absence of large maria. Of all the main near-side maria, only the Mare Orientale extends far onto the averted regions, and there is nothing on the far side even remotely comparable with, say, the Mare Imbrium. Even the so-called Mare Moscoviense with a diameter of about 420 km is relatively small. There are craters everywhere, and even basins of mare size that are light-floored rather than filled with dark mare material. There seems no reasonable doubt that this difference is due to the fact that the lunar rotation has been synchronous since a fairly early stage in the story of the Earth–Moon system.

Hot spots
During lunar eclipses (*see* pages 10–11) there is an abrupt cooling of the Moon's surface, and this fact has proved useful to astronomers investigating the Moon's geology from Earth. Since the Moon is without atmosphere and the surface materials are very poor at retaining heat, the sudden fall in temperature is quite dramatic. Yet not all areas cool down at the same rate. Infrared techniques have revealed that there are some regions which cool much less rapidly than their surroundings, and these have become known as "hot spots"; the best example is the great ray-crater Tycho, in the southern uplands of the Moon. These so-called hot spots are also warmer than their environs during the long lunar nights. There is no suggestion that internal heat is responsible; the effect is due solely to a difference in the nature of the surface materials.

From time to time there have been suggestions that the sudden cold may produce visible effects upon certain special features of the surface, notably the craterlet Linné on the grey plain of the Mare Serenitatis, which is surrounded by a white nimbus. However, these results are extremely dubious, and it is not likely that any real effect occurs. Nor have major structural changes occurred on the Moon for many millions of years.

The Moon's Interior

The mean density of the lunar globe is 3.34 times that of water, a value which is much lower than those of the Earth (5.5), Mercury (5.4), Venus (5.2) or even Mars (3.9). It is therefore reasonable to assume that there is no large, heavy, iron-rich core comparable with that of Earth, although magnetics investigators argue that there must be a small core to explain.

It is important to establish first of all whether the Moon is hot inside, or whether it is a cold globe throughout. Experiments designed to measure the heat outflow were among the most important of those carried out by the later Apollo missions, but unfortunately they were not entirely successful. Instruments were carried on board Apollos 15, 16 and 17, but that of Apollo 16 was broken while being set up, and attempts to carry out makeshift repairs proved to be abortive. However, the results from the other two experiments were conclusive. Apollo 15 showed that the outflow is about half that of the Earth—the temperature rose by 1 degree K at a depth of 2 m—and the experiments from Apollo 17 gave results of the same order. The "cold Moon" theory was at once abandoned, and it became clear that the interior is at a high temperature. The lunar heat flow may be accounted for by the presence of the radioactive elements potassium, uranium and thorium (*see* pages 28–29).

Moonquakes

In the terrestrial globe earthquakes produce waves of various types, some of which can travel through molten or liquid material while others cannot; research into the behavior of such waves has provided most of the information about the Earth's interior, and it was hoped that the same would be true of the Moon. Each Apollo ALSEP, therefore, included a seismometer or "moonquake recorder". It was found that there are three distinct kinds of moonquake. The first kind are those produced by the impacts of meteorites, the second are those produced artificially, either by exploding small charges on the surface during Apollo missions or (more importantly) by crashing the IVB ascent stage of the Saturn rocket, or the Lunar Modules after the astronauts have rejoined the orbiting section of Apollo, and the third are due to internal movements in the lunar globe.

Of the meteoritic impacts, the most important so far detected occurred in July 1972; the object is thought to have weighed about 1,000 kg. Meanwhile, the Module impacts produced unexpected results. It was found that the vibrations set up did not die abruptly away; they lasted for over an hour, so that the materials below the crust did not immediately dampen the shock.

The third group, natural moonquakes, has proved to be the most valuable. Natural moonquakes may be produced in a variety of ways: for example, deep moonquakes, which originate at depths of 600 to 950 km, are thought to be of tidal origin, while shallow moonquakes may be caused by the expansion and contraction of the surface rock under the influence of solar heating. Moonquakes had been predicted long before the Apollo period, mainly because observers had detected minor glows and local obscurations on the surface—known today as TLP or Transient Lunar Phenomena—which were attributed to minor disturbances; it was found that they were commonest in certain specific areas, notably that of Aristarchus, and that they were most common near perigee, when the lunar crust was under maximum gravitational strain. Suggestions that TLP would be associated with moonquakes proved to be well-founded.

There were found to be less than 3,000 moonquakes per year on average, and they never exceed a value of 2 on the Richter scale, which means that by terrestrial standards they are very mild; the total yearly energy released is only about 2×10^6 J, against 10^{17} to 10^{18} J for Earth (they will present no hazards to future lunar bases). But they are evident enough, and they have led on to a picture of the Moon's interior which is probably reasonably accurate.

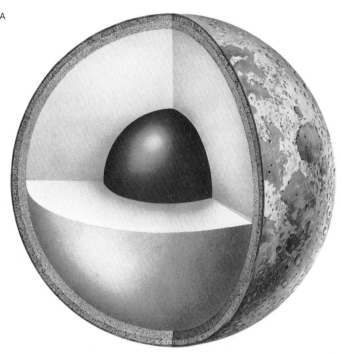

1. Interior
The structure of the Moon (**A**) differs markedly from that of the Earth (**B**). Both crust and mantle are thicker on the Moon, no doubt because the internal temperatures are much lower. The upper regolith, made up of shattered bedrock, is relatively shallow; below comes the crust, extending down to 60 km. The upper crust (to 20 km) is more solid rock; the remainder is made up of materials with properties similar to those of the anorthosites and feldspar-rich gabbros of the highlands. Below the crust lies the mantle, and below this again is the asthenosphere, assumed to be a region of partial melting. Finally there is the relatively dense core.

2. Moonquake waves
Studies of natural moonquakes have yielded considerable information about the lunar interior. At 20 km there is a sharp increase in the wave velocity to 7 km s⁻¹, which remains fairly constant down to a depth of 60 km, indicating that most of the upper cracks have been filled in. There is an abrupt change at 60 km, the bottom of the crust. Useful information also came from the impact of a 1,000 kg meteorite on 17 July 1972; the waves set up indicated that below 1,000 to 1,200 km in depth the lunar rocks are hot enough to be molten.

3. Seismographic traces
Lunar seismic signals (**A**, **B** and **C**) take much longer to die away than those of the Earth (**D**).

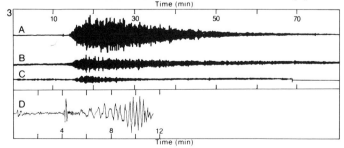

B

4. Mascons, epicenters and TLP sites
Mascons were discovered from the orbital movements of probes circling the Moon. Mascons are associated with circular maria such as Serenitatis and Imbrium, and smaller formations of the same basic type such as Grimaldi. They indicate differences in the thickness or composition of the crust in these regions as compared with their surroundings. Also shown are the main epicenters of moonquake regions and regions in which red TLP have been reliably reported. It is clear that the distribution of these three types of features is not entirely random.

4 Mascons

★ Deep seismic epicentres

● TLP sites

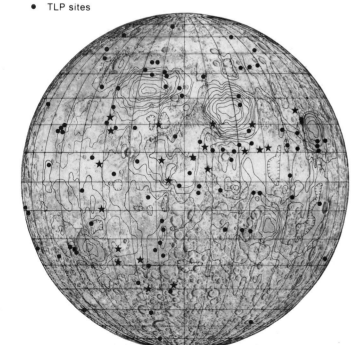

Structure

The velocities of moonquake waves (both natural and artificial) give useful information about the Moon's layered structure. Down to 1 km in depth the compressional waves have velocities increasing from $100\,\text{m s}^{-1}$ to $900\,\text{m s}^{-1}$. This indicates a change from the "regolith", the outermost layer of the Moon, to firmer and more consolidated material. By a depth of 20 km the velocities reach $6\,\text{km s}^{-1}$, indicating fractured basaltic material (*see* pages 22–23), which increases in density with depth. At 20 km there is a velocity increase to $7\,\text{km s}^{-1}$, and this remains constant down to 60 km depth, which agrees well with the idea of anorthositic gabbro composition, which has properties similar to the surface highlands. Below 60 km the velocity goes up to $9\,\text{km s}^{-1}$, but the data are not yet reliable. This marks the limit of the Moon's crust. The lower new rocks, rich in the substances known as pyroxene and olivine, are relatively dense. They go down to 150 km. Beneath this is a region extending down to 1,000 km, termed the lithosphere; it is solid and rigid, with the main moonquakes occurring near its base. Below 1,000 km the shear waves are weakened; this is the asthenosphere, which may indicate the presence of a liquid or at least partially melted core with a diameter of from 1,200 to 1,800 km (as was shown from studies of the meteoritic impact of July 1972).

Finally there is the true core, the composition of which is uncertain. If it were pure iron, it could not have a diameter greater than 1,000 km, because it would have a detectable effect on the Moon's motion, and because of the high melting point of iron—1,808 K—it would be hard for such a core to form and remain molten. However, if the core contained some iron sulphide (FeS), the melting point would be only about 1,300 K and the entire core might be up to 1,400 km in diameter. It must be admitted, however, that as yet knowledge of the Moon's interior is very incomplete; the best hope of obtaining more reliable information will be if by chance a sufficiently large meteorite falls in the right place while a seismometer network is operating on the Moon.

It is interesting to note that the crustal thickness is greater on the Moon's far side than on the visible hemisphere and may go down to over 70 km. This again must be due to the fact that the rotation has been synchronous for so long a period.

Of great importance are the mascons, which appear to be regions of denser material lying not far below the Moon's surface. They were first detected by studies of the movements of the US probe Orbiter 5. If a probe passes over a region where the lunar material is unusually dense, it will speed up slightly (the effect is known as a "positive gravity anomaly"); if it passes over an area where the material is less dense, it will slow down ("negative gravity anomaly"). This was the case with Orbiter 5. Positive anomalies were found with some of the regular maria—Imbrium, Serenitatis, Crisium, Nectaris, Humorum, Humboldtianum, Orientale, Smythii, Aestuum and the walled plain Grimaldi, which is very dark floored and presumably of the same nature. One explanation is that they were produced by the transformation of lunar basalts to denser rock at the edges of the circular formations. On the other hand, craters of the Copernicus type tend to show negative gravity anomalies, and so do the large unfilled basins, most of which lie on the far side of the Moon.

Finally, what of the Moon's magnetic field? Today the overall field is negligible; this was established as long ago as 1959 by the second Luna probe. Yet there are some regions which show traces of magnetic fields here and there—for instance, at the small crater named North Ray Crater, studied by the astronauts of Apollo 16. There is also a local field near the formation Van de Graaff, on the far side, which is a depression 4 km deep and is slightly richer in radioactive materials than its surroundings. The evidence from paleomagnetic studies of rocks indicates that there used to be an appreciable overall magnetic field around 3,000 million years ago, which has now disappeared.

The Regolith

In contrast with most of the planets, whose colors are striking, the Moon's appearance gives the impression of an almost complete absence of local color. Mars has its red deserts and its darker areas, Saturn is yellow, while Uranus is decidedly green, and the overall yellow hue of Jupiter is modified by features such as the Great Red Spot; but on the Moon there is nothing to relieve the greyness of the plains except the blackness of the shadows and brighter areas of the lunar highlands.

The descriptions given by the Apollo astronauts fully confirm this impression. Charles Conrad, commander of Apollo 12—which came down in the Mare Nubium—commented that "the Moon is just sort of very light concrete color. In fact, if I wanted to look at something I thought was the same color as the Moon, I'd go out and look at my driveway."

The surface of the Moon is covered by what is termed the "regolith", which may be defined as a debris blanket made up of loose material of rock fragments and "soil", overlying solid bedrock. Conventionally, soils are defined as particles less than 1 cm in diameter, while larger particles are termed rocks. The thickness of the regolith varies considerably. Direct measurements have been made only from the sites of the successful Apollo missions (11, 12 and 14 to 17 inclusive)—a mere half-dozen in all—but they were obtained from diverse areas and provide reliable clues. It seems that over the maria the average depth of the regolith is from 4 to 5 m, while over the highlands it may go down to 10 m and even deeper in places—perhaps as much as 30 m, although this may be rather exceptional. On the other hand, the Apollo 16 astronauts found that at North Ray Crater the regolith went down to a depth of only a few centimeters.

Because the regolith is loose, it retains impressions. Pictures of the prints left by the astronauts have been widely published, and the prints left by the footpads of space vehicles that have landed provide similar evidence of the fine, powdery quality of the lunar surface. The prints themselves will remain visible for a very long time before they are eventually obliterated. Yet despite the lack of atmosphere, which means that there is no wind and indeed no "weather" on the Moon, the regolith does not stay undisturbed indefinitely. In particular, there is a constant bombardment from space. Meteorites can reach the lunar surface without being destroyed (as happens over the Earth—there can be few people who are not familiar with "shooting-stars", which are, in fact, meteors or burnt-up meteorites) and so can the much smaller and more numerous micrometeorites. In fact the regolith is churned up, although the effects are very slow and would be quite undetectable over a period of many human lifetimes. The regolith has itself been produced by this kind of external action, and there has been plenty of time for this to happen, since the lunar surface solidified so long ago. The age of the highland regolith is of the order of 4,000 million years.

Composition

A few terms must be introduced here. "Magma" is molten subsurface rock, which becomes "igneous rock" when it solidifies and "lava" when it comes out onto the surface. "Basalts" are fine-grained, dark, igneous rocks. They are composed chiefly of plagioclase feldspar and pyroxene with or without olivine. "Anorthosite" is an igneous rock made up chiefly of plagioclase; some anorthosites may indeed be almost pure plagioclase. They are relatively rare on Earth and are found mainly in Precambrian rocks such as those of Quebec and Labrador, but they are common on the Moon, although not identical in composition. "Pyroxene" is a calcium–magnesium–iron silicate and is the most common mineral in lunar mare rocks, making up about half of most specimens; it forms yellowish-brown crystals. "Olivine", a magnesium–iron silicate, is also found and is characterized by pale green crystals. "Gabbro" is a coarse-grained igneous rock, dark in color; it may be considered to be a coarse-grained equivalent of

2. Lunar Rock Exteriors
This montage shows a selection of lunar rock exteriors. (**A**) The fine-grained crystalline texture is typical of a mare basalt. Both dark and light minerals are distinguishable. (**B**) Mare basalts can also have this highly "vesicular" appearance—the vesicles were formed by the release of gas as the rock cooled. The nature of the gas is, however, unknown. (**C**) Highland basalts are almost entirely plagioclase and therefore are almost completely white in color. The dark patch on this sample is splashed impact-produced glass. Zap pits produced by micrometeorite impacts are also visible on the lower central area of the rock (see 5**A**). (**D**) A typical "polymict" breccia, consisting of larger fragments set in a fine-grained matrix; this is

not a primary rock, its constituents were formerly components of other materials.

3. Lunar soils
In addition to single mineral grains, which are in fact comparatively infrequent, igneous rock fragments, breccias and glass spherules (**A**) abound in lunar soils. These objects, which can be either round or dumbell shaped, are mainly the result of impact-produced glass that solidified in flight. When the glass has landed back on the surface before solidification, "glassy agglutinates" are formed (**B**). These particles, which can account for up to 60 percent of a lunar soil, consist of the whole spectrum of soil components held together by a fragile web or matrix of splashed glass.

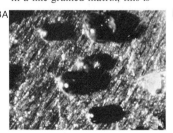

1. Lunar surface
The fine texture of the regolith can be seen clearly in the impression made by the footpad of Surveyor III. The vehicle bounced slightly on touchdown, leaving two footprints before it finally came to rest.

4. Lunar rock interiors
The minerology of rocks can best be studied after samples have been cut into thin slices and polished. A typical mare basalt (**A**) is revealed as being mainly pyroxene (grey), plagioclase (white), and ilmenite (black) by microscopic study. (**B**) When viewed in polarized light plagioclase appears black and white striped. This rock, the so-called "Genesis Rock", is an anorthosite made up almost entirely of plagioclase, with small amounts of pyroxene. (**C**) Microscopic study also reveals the different textures of the rocks. Compare the gabbro shown here (pyroxene, dark) with the basalt in (4**A**). The crystals' shapes are clearly different and thus are said to have different "habits". (**D**) This photomicrograph exhibits a breccia texture—larger fragments are imbedded in a finer-grained matrix. The rock is not a typical breccia, because only one mineral (olivine) is present, but is called a "dunite". A sample from this specimen yielded a radiometric age of 4,600 million years, nearly the age of the Moon.

4A

B

0·5 mm

0·5 mm

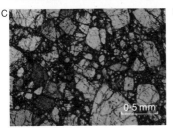

C

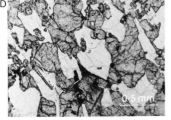

D

0·5 mm

0·5 mm

5. Lunar rocks and soil grains
The scanning electron microscope reveals a crater on a glassy spherule like one of those in (3**A**), showing that micrometeorites extend the range of impacts on the Moon over a scale of many orders of magnitude. (**B**) In this transmission electron micrograph, electron diffraction demonstrates that the interior of a particle remains crystalline while the exterior has been rendered amorphous by intense solar wind bombardment. (**C**) Cosmic ray particles also leave evidence or "tracks" of their presence in the lunar soil; these pits were produced by etchng material that has been damaged by radiation. (**D**) Solar wind bombardment also causes microscopic chemical changes—these dark spherules are iron formed in this way.

5A

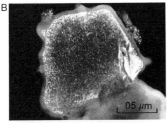

B

10 μm

05 μm

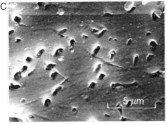

C

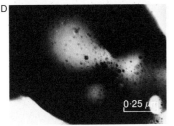

D

5 μm

0·25 μ

basalt. Lunar basalts and gabbros may also contain ilmenite, an opaque iron–titanium oxide. "Breccias" are fragments made up of shattered, crushed and sometimes melted pieces of rocks; these pieces may be igneous rocks of various types, glasses, or bits of other breccias. It is worth noting that breccias have been collected from every landing site on the Moon. There are, for instance, widespread layers of breccias at the sites of Apollo 14 (Fra Mauro) and Apollo 16 (the so-called Cayley Formation, in the region of Descartes). One mineral first found on the Moon has been named "Armalcolite" (after the three astronauts of Apollo 11— *Arm*strong, *Al*drin and *Col*lins): it is an opaque oxide of iron, titanium and magnesium. It is reasonably common, particularly in the rocks returned from the Taurus–Littrow area, and has been recognized in terrestrial rocks since it was identified in the Apollo 11 samples.

No entirely new substances have been found on the Moon. All lunar material is made up of the elements familiar on Earth, though there are modifications in composition due to the fact that the materials have developed under conditions which have been completely different.

The regolith is not the same over the maria as over the highlands. On the maria the rock fragments are chiefly basalt, with pyroxene, plagioclase and other minerals. On the highlands the fragments are chiefly plagioclase-rich rocks and broken plagioclase crystals. As yet there is no direct information about the regolith on the Moon's far side (at least, there are no samples of it), but it is not likely to be substantially different from that on the Earth-turned hemisphere.

Glasses are very common in the regolith, Much of the soil is composed of tiny rock fragments held in a matrix of glass—such particles are termed agglutinates. Occasionally discrete glass beads of various colors are found, such as, for example, emerald-green glasses found in the Apollo 15 soils which are rich in calcium. These beads are formed when glass which has been melted—usually as the result of impacts—re-solidifies in flight. The resulting particles can be either spherical or dumbell shaped. One startling discovery was that of the "orange soi'" at Shorty Crater in the Apollo 17 site. It was found to be almost entirely tiny colored glass particles, from 0.1 to 0.2 mm in diameter, and to be no younger than the surrounding material; the age has been given as 3,800 million years. One ingenious attempt to explain the abundance and uniformity of composition is that the orange glass was formed by an impact into a lava lake, and was subsequently excavated by Shorty Crater. Samples of soil brought back as cores up to 2.5 m in length show distinct layering, and demonstate how the regolith has developed as a result of a sequence of impacts.

Many samples from the regolith have been brought back for analysis, and all have interesting features of their own. For example, there is the so-called "Genesis rock" from Apollo 15, which is an anorthosite 4,000 million years old, but still not the oldest lunar rock; and from Apollo 12 comes Sample 12013, which is lemon sized and made up of a dark grey breccia, a light grey breccia and a vein of solidified lava. Its dark portions contain much more than the usual quantity of potassium, as well as having an increased proportion of uranium and thorium, so that it is the most radioactive rock known from the Moon.

There is a certain amount of meteorite material (about 1 to 2 percent) in the regolith, although certainly not nearly as much as some authorities had expected. Finally, it had been suggested that tektites might come from the Moon—tektites are aerodynamically shaped objects found in restricted fields on the Earth, notably in Australia, and their origin is a completely mystery. They are usually button sized and are unlike any other objects found. However, their chemical composition is so unlike that of known lunar materials that whatever they may be, they did not come from the Moon, either shot out from lunar volcanoes or hurled earthward by meteoritic impacts.

Lunar Chronology

When discussing the evolution of the lunar surface, cartographers have often described features as "young" or "old", deriving relative ages or "stratigraphy" on the basis of whether one formation overlies another. However, because the American and Russian missions returned samples from the Moon, it has been possible in recent years to place an absolute time scale on some of the events which have occurred in lunar history. The key to this understanding has been the measurement of the relative abundance of certain long-lived radioactive isotopes such as uranium, rubidium and potassium and their daughter products (appropriate isotopes of lead, strontium and argon) which date the moment at which the rocks were last strongly heated or possibly even molten. "Young" by lunar standards turns out to be extremely ancient in the terrestrial sense. It is not known for certain when higher forms of life started to develop on Earth, but 600 million years ago may be a reasonable estimate. By then, almost all the familiar structures on the Moon's surface had been produced.

Strangely, more is probably known about the very early history of the lunar surface than about that of the Earth. The Moon has been almost static for well over 3,000 million years, whereas the Earth is subject to change all the time: the outlines of the seas and continents alter, old mountains are eroded away while others rise up, and there is continuous volcanism. As a result, the earliest evidence has been obliterated, so that knowledge of what the Earth must have looked like 4,000 million years ago is very sketchy indeed. The Greenland metasedimentary rocks, which are among the oldest surviving terrestrial samples, date back less than 3,800 million years, whereas most of the lunar highlands (85 percent of the Moon) are considerably older.

The Moon and the Earth are believed with reasonable certainty to be approximately the same age, about 4,600 million years old. It is also generally assumed that they were produced by accretion from material in the so-called "solar nebula", a cloud of particles and gases associated with the primeval Sun. The process generated considerable heat, and the outer layers of the newly formed Moon were probably molten down to a depth of several hundred kilometers; some less dense minerals separated out to the surface and produced a primitive lunar crust as the Moon slowly cooled.

The next period saw the production of the huge ringed basins which now are occupied by the maria. Not all are the same age. When the impacts that created these basins occurred, the intense heating that resulted had the effect of resetting the radiometric "clocks": from the Apollo samples of ejecta produced by the basin-forming events, some formation ages are known with certainty. Thus, the Imbrium basin was formed close to 3,900 million years ago. Orientale is slightly younger at perhaps 3,850 million years, whereas Serenitatis has been dated at over 4,000 million years. Considerably older basins also exist. Tranquillitatis and Fecunditatis may be 4,500 million years old. These greater ages, however, have not been derived by direct radiometric dating but are really relative ages based on such considerations as the way the features seem to fit into the sequence of events and the state of erosion of their associated rings. The generally accepted order of basin formation is now taken to be: oldest—Tranquillitatis, Fecunditatis, Nubium; intermediate—Serenitatis, Nectaris, Humorum; and youngest—Crisium, Imbrium and Orientale.

When studying samples of lunar highland rocks, chronologists have been struck by the clustering of ages around 3,900–4,000 million years. It seems that during the first 600 million years of its life the Moon experienced a bombardment far more severe than has occurred subsequently. It is not known whether the early bombardment, or the lunar cataclysm as it has been called, was a slowly declining process or whether it was a sudden climax signalling the last stages of the accretion. What is certain is that this period has all but obliterated the primitive lunar crust; only the faintest trace can be found today as isolated inclusions in complicated breccias (*see*

1. 4,600 million years
Pre-Imbrian period: formation of the Moon; melting and separation of the crust by the process of "magnetic differentiation", which produced the lunar highlands (*see* pages 28–29). As the molten material cools, it begins to crystallize, and the lower-density crystals tend to rise to the surface. Some material as old as 4,600 million years has been found, but most of the original crust has been destroyed by cratering.

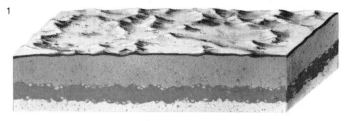

2. 4,500 to 3,850 million years
Formation of the great mare basins (*see* pages 26–27) during a period of cataclysmic bombardment. The Tranquillitatis basin may be as old as 4,500 million years; Orientale, only 3,850 million years.

3. 3,900 to 3,200 million years
Imbrian period: the mare basins were flooded with lava, which poured out of the interior of the Moon in successive stages. These formed layers of mare material and left the maria looking much as they do today.

4. 3,200 to 1,000 million years
Eratosthenian period: following the rather abrupt end of the flooding period, crater formation was restricted to the effects of minor volcanism and impacting meteorites. Copernicus may be about 1,000 million years old.

5. 1,000 million years to the present
Copernican period: the Moon's recent history is characterized by almost complete quiescence, both volcanic and meteoritic, although a few of the ray-craters may date back less than 1,000 million years.

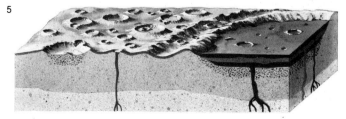

6. Reconstruction of the Moon
This sequence shows the Moon as it is believed to have appeared (**A**) in the middle of the Imbrian period after the formation of the last of the mare basins, before they had been flooded; (**B**) at the end of the Imbrian period, soon after the formation of the mare material approximately 3,300 million years ago; and (**C**) as it appears at the present time (according to telescopic photographs made in 1966).

6A

B

C

pages 22–23). It has often been noted that the earliest rocks found on Earth just post-date the end of the lunar cataclysm—whether this is a coincidence or a significant observation is not yet known.

At the end of the cataclysm the surface of the Moon probably looked much like diagram 6A. The end of this phase, however, denoted the start of the next period of lunar evolution—the age of tremendous volcanic activity which lasted from 3,900 to 3,200 million years ago. For extended periods, layer by layer, lavas flooded into the great basins from about 100 km beneath the lunar crust. Even to the most casual observer the maria are younger than the highlands. A cursory examination reveals the low relative crater density on mare surfaces. For the regions from which samples have been returned the ages of the last basalt flows are very well known. Thus, Tranquillitatis (3,800–3,600 million years) was flooded before Fecunditatis (3,400 million years), which preceded Oceanus Procellarum (3,200 million years). It appears that there is some relation with chemistry in that basalts high in aluminium are slightly older than those high in titanium (rich in the mineral ilmenite); those rich in the mineral olivine were among the last to crystallize. Since the ages of the mare basalts are far removed from the ages of the basins there is no question about the heat source generating the lava: it could not have arisen from the impacts which produced the basins, but must have come from the decay of radioactive uranium, thorium and to some extent potassium.

For almost 700 million years lava poured out from the Moon's interior, flooding the mare basins and producing the great dark plains which can be seen today. Many of the basins were joined by these lava-flows: thus the boundary between the Mare Serenitatis and the Mare Tranquillitatis has been breached, while the Mare Imbrium, the most prominent of all the regular maria, joins on to the Oceanus Procellarum and the Mare Nubium. There is also a break in the boundary between the Apennines and the Caucasus ranges leading into the Mare Serenitatis. Most of the basins on the lunar far side, however, remain unfilled. The lava-flows ended about 3,200 million years ago, probably rather abruptly. The Moon was starting to take on its present appearance, and even at that remote period the outlines of the great plains such as the Mare Imbrium would have been recognizable (*see* diagram 6B).

Meanwhile, the walled formations were being produced, both the very large structures such as Bailly, Grimaldi and Ptolemaeus, and smaller craters with central peaks. Crater formation continued long after the lava-seas had become permanently rigid, and it may be assumed that the youngest are those which are the focal points of bright ray systems, such as Copernicus, Tycho, Kepler and Olbers. Copernicus may be no more than 1,000 million years old, and Tycho perhaps even less. Yet even at the time of the formation of Tycho, dinosaurs on the Earth still lay in the far future.

As far as the relative ages of lunar features are concerned, much can be learned from seeing how they are arranged. For instance, in the great chain of walled plains on the boundary of the Mare Nectaris there are three main formations: Theophilus, Cyrillus and Catharina. Theophilus overlaps Cyrillus, so there can be no reasonable doubt that it is the younger of the two. Similarly, the ruined walled plain Janssen must be older than Fabricius, which interrupts it. The ray-craters are particularly informative. The fact that the rays overlie all other formations indicates that they are among the most recent of features. Crater frequency counting also helps to decide relative ages.

Since the formation of ray-craters, the Moon has been relatively quiescent; it is most unlikely that any of the large formations are more recent. However, there are plenty of smaller features that are relatively youthful. Some have been reliably dated by measuring exposure ages of ejecta (the length of time material has been exposed to solar particle fluxes). Thus, Cone Crater, visited by Apollo 14 astronauts, is probably only 20 million years old; North Ray at the Descartes site possibly 60 million years.

The Maria

The grey plains or "maria" occupy about 15 percent of the Moon's surface. They are the most obvious of all the lunar features and are easily visible with the naked eye. The maria are of two distinct types, regular and irregular; the regular maria being basically circular depressions with mountainous borders that are frequently incomplete. The maria on the Earth-turned face of the Moon lie between 2 and 5 km below the mean radius of the globe.

The most prominent of all is the Mare Imbrium, whose formation had profound effects over much of the Moon's surface. It is surrounded by mountains, although in places the border has been obliterated, and there are indications of an inner ring whose southern boundary is traced by the craters Archimedes, Timocharis, Lambert and C. Herschel. Studies of the overall arrangement are most significant. It seems that lunar "mountain ranges" are not of the same type as those found on Earth, but are in fact the "walls" of the regular maria or the uplift caused by cratering; the Mare Imbrium, for example, is surrounded by the Apennines, Alps and Caucasus, while the adjoining Mare Serenitatis is bounded in part by the Caucasus and Haemus ranges.

Most of the main maria are connected, forming one huge system, although the separate basins are of different ages and have different depths. The chief exceptions are the Mare Crisium, in the northeast quadrant of the visible hemisphere, and the Mare Orientale, which lies very close to the western limb and actually extends onto the far side; other separate seas are the Mare Humboldtianum and the Mare Smythii.

Other maria are less regular in outline and are certainly older than the Mare Imbrium. Examples include the Mare Nubium, Mare Tranquillitatis, Mare Fecunditatis and Oceanus Procellarum, which are independent basins but have been less well preserved. The genuinely irregular maria are quite distinct, and some appear to be mere coatings of dark material. The best example is the Mare Frigoris, which has no definite shape at all, while the Mare Australe, in the south-east quadrant close to the limb, and the Mare Undarum, close to the Mare Crisium, do not appear to be true basins.

It is difficult to draw a distinction between a "mare" and a large "walled plain". Grimaldi, for example, is a vast low-walled enclosure, more or less circular, with a diameter of almost 200 km. It has a dark floor of marial type, and if it had been placed closer to the center of the disc, there is every possibility that it would have been classed as a "mare" rather than a "crater"; its diameter is not much less than half that of the Mare Crisium. Thus Grimaldi and the Mare Crisium may well be of the same fundamental type, a conclusion reinforced by the fact that both are associated with mascons (*see* pages 20–21). There are other walled plains with mare-type interiors, such as Plato, for instance, with a diameter of nearly 100 km. Many authorities doubt whether there is any essential distinction between a large, dark-floored walled plain and a regular mare, except in size. It is also worth noting that Grimaldi is the darkest patch anywhere on the visible side of the Moon, and the interior of Plato is comparable.

Then there are the "lakes" and "marshes", such as the curiously coloured Palus Somnii, which adjoins the Mare Tranquillitatis and is bounded by two of the rays issuing from the brilliant crater Proclus. Most spectacular of all is the Sinus Iridum or "Bay of Rainbows", which leads out of the Mare Imbrium. It is bordered on its limbward side by moderately high mountains, and under suitable conditions of illumination it seems to project beyond the terminator; it has been nicknamed the "Jewelled Handle". There are traces of an old border between it and the Mare Imbrium, and since it is circular in form (although foreshortened into an ellipse as seen from Earth) it is presumably a separate basin.

Some of the maria are surrounded by external mountain arcs. One such case is the Mare Nectaris. The feature once known as the Altai Range, and now more appropriately known as the Altai

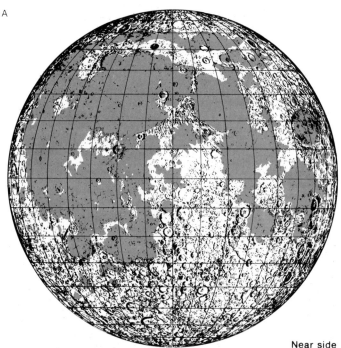

1A

Near side

B

Far side

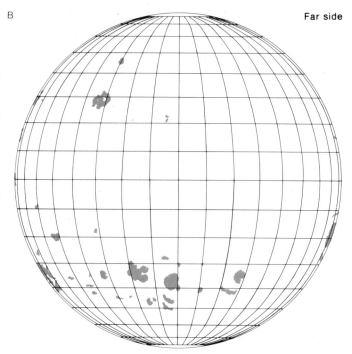

1. Distribution of lunar maria
A large proportion of the Moon's near side (**A**) is occupied by maria, while the far side (**B**) is notably lacking. The reason for this difference is thought to be associated with the Earth's gravitational effect on the Moon, but the mechanism involved is not yet understood. The diagram shows how many of the near-side maria flow into one another, forming the familiar pattern clearly visible from Earth. The regularly shaped maria are surrounded by mountainous walls.

Chemical composition of basalt samples (weight%)

	Earth Ocean-floor	Moon Maria	
SiO_2	49.2	37.0	– 49.0
TiO_2	1.4	0.3	– 13.0
Al_2O_3	15.8	7.0	– 14.0
FeO	9.4	18.0	– 23.0
MnO	0.16	0.21	– 0.29
MgO	8.5	6.0	– 17.0
CaO	11.1	8.0	– 12.0
Na_2O	2.7	0.1	– 0.5
K_2O	0.26	0.02	– 0.3
P_2O_5	0.15	0.03	– 0.18
Cr_2O_3		0.12	– 0.70

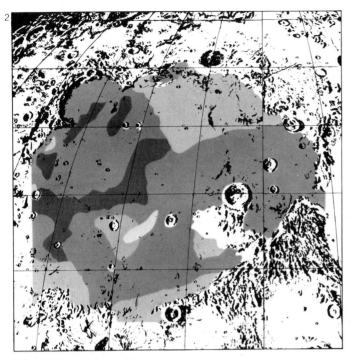

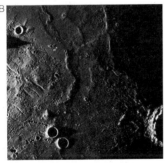

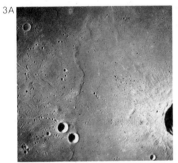

2. Imbrium lava flows
Differences of color and crater density in the Mare Imbrium have revealed a difference in ages of the lavas. In the diagram the darkest shading represents the most recent lava flows.

3. Mare Imbrium
Under different lighting conditions, the boundaries between geological units appear to be different, as in these contrasting pictures taken when the angle of the sun was 17° (**A**) and when it was 2° (**B**).

4. Typical mare features
The maria (**2**) differ from the highland regions (**1**) in being smoother and generally darker in color. They are marked by a variety of features, such as sinuous rills (**3**), wrinkled ridges (**4**) and dome-like features with summit craters (**5**). Sinuous rills are interpreted as lava channels, often originating from craters in relatively high regions of maria and flowing into lower lying areas. Wrinkle ridges are believed to result from folding in the surface layers of the maria.

Scarp, is concentric with the mare border and was certainly produced by the same event. The Mare Orientale, meanwhile, has several rings, some of which are visible from Earth and have been named. The circular outer scarp—the Cordillera Mountains—is over 900 km in diameter and is extremely massive; within this ring lie the Rook Mountains, making up another circular scarp almost 650 km in diameter.

On the Moon's far side the whole situation is different. Apart from a section of Orientale, there are two features that are known as maria—Moscoviense and Ingenii—but neither really qualifies. Mare Moscoviense is dark floored and was recognizable on the very first pictures of the far side sent back by Luna 3 in 1959, but it is smaller than some of the other far-side basins such as Apollo and Hertzsprung, while the so-called Mare Ingenii is merely an irregular darkish area near the interesting walled formation Van de Graaff.

Crater frequency on the maria is, of course, noticeably less than in the highlands, and it is reasonable to assume that most of the marial craters are fairly young by lunar standards. Copernicus is a striking example. Not far to the east is the famous "ghost", Stadius, which is 70 km in diameter, with walls that are barely traceable; it is pitted with small craterlets. Ghost-craters are found on most of the maria, and in many cases the so-called "wrinkle-ridges" are in fact parts of the "walls" of such features, though other wrinkle-ridges are probably compressional features. There are prominent wrinkle-ridges on the Mare Serenitatis, but very few craters; the largest of them, Bessel, is only 19 km in diameter.

There are craters belonging to the basin-forming era whose "seaward" walls have been destroyed by lava-flows, so that they have become bays. The Sinus Iridum comes into this category, and another excellent example is Fracastorius, in the south of the Mare Nectaris, where the old wall can just be traced, but is of negligible height. Doppelmayer and Hippalus, leading off the Mare Humorum, have been similarly damaged, though each has the remains of a central peak, and Hippalus is associated with a major system of rills.

Composition
So far, specimens have been obtained from only a few lunar sites, but it has been found that these specimens are not identical—there are local variations, which is by no means unexpected. The rocks found in the maria are of basaltic origin—as are all the lunar rocks. They differ from the highland basalts in that they may be obtained as distinct samples rather than merely as components of complicated breccias. Their nearest terrestrial counterparts are the lavas that are continuously erupting from the spreading ocean ridges, but there are very marked chemical differences (*see* Table). The lunar mare basalts contain much less silica (SiO_2) and aluminium but far greater amounts of iron, magnesium and titanium. The local differences between rocks are reflected in the relative proportions of these elements. Apollos 11 and 17 sampled areas where titanium content was very high—near Hadley Rill the rocks were rich in magnesium and silica. Volatile elements such as potassium and sodium are also depleted on the Moon. Likewise neither mare nor highland rocks contain water or even traces of hydrated minerals. They must have been formed under extremely dry conditions. Any grandiose idea of extracting water for the benefit of future colonists has had to be abandoned. Moreover, since aqueous geologic processes are frequently involved in the concentration on Earth of economic deposits of important metals such as copper, there now seems little likelihood of the Moon becoming a future source of raw materials.

One of the consequences of the unusual chemistry of lunar mare rocks is a very low viscosity, hence the scale of lunar lava flows. Rocks of this fluidity were not known before the lunar missions and this ignorance contributed to the difficulty of accepting that the maria were not some kind of sedimentary formation.

The Highlands

The highlands or "terrae" are the most ancient parts of the lunar surface. They were the first to solidify, and were not penetrated by the tremendous volcanic flows that flooded the mare basins. The highlands were produced by the process known as "magmatic differentiation". Materials in an igneous melt separate chiefly according to their densities: as the lighter materials crystallize they float to the top, forming lower-density regions. Measurements carried out largely by the orbiting sections and the sub-satellites of the Apollo vehicles have confirmed that the density in the highlands is indeed less: it averages 2.75 to 3 times that of water, as against a value of 3.3 to 3.4 for the maria. The highland crust appears to be in a state of "isostatic equilibrium" (that is to say, in elevated regions the material tends to be less dense than in depressions, so that the total weight pressing down is much the same everywhere).

But although the highland surface is ancient even by lunar standards, it has not remained completely undisturbed since it became solid. The cataclysmic bombardment of highlands before 3,900 years ago has severely brecciated almost all the highland rocks so that remnants of the primitive crust may only be found disseminated in such samples. A veil has been drawn over the early history of the highlands of the Moon; it will only be disclosed by painstaking effort. There are also indications of tectonic activity: for instance there are many faults, or pairs of faults close together, bordering grabens. Rills are less well-marked than in the marial regions, and most of the large so-called valleys, such as those of Rheita and Teichenbach, turn out to be chains of confluent craters; there are also small "strings of beads", such as the 225 km chain near the crater Abulfeda, which can only be volcanic. On a smaller scale, too, there are visible tracks of boulders that have rolled down slopes and left their impressions, probably because they have been jolted by sudden crustal movements. There was one excellent example of this in the area of North Massif, at the Apollo 17 landing site, and there is another inside the crater Vitello on the edge of the Mare Humorum, so that such events are not confined to the terrae.

Before the Apollo and Luna missions the highlands were less well explored than the maria, because the features are so crowded; there are craters in profusion, and the amount of detail to be seen is daunting to even the most enthusiastic telescopic observer. It was evident that landing in a rough region presented more problems than coming down in one of the smoother plains, and only one of the manned missions was aimed at the highlands. This was Apollo 16, which touched down in the so-called Cayley Formation in the region of Descartes. It was indeed unfortunate that the heat-flow experiment failed and that the data in this field of research remain incomplete. However, it has been established that magnetic fields found locally in the highlands are stronger and less uniform than in the maria.

Composition

The highland rocks, although still basalts, are not identical in composition with those of the maria, and are certainly very different from the terrae or continental rocks on Earth, of which granite is an example (see Table).

On the basis of tiny chips found in the Apollo 11 lunar soils obtained from the Mare Tranquillitatis, it was postulated that the lunar highlands would be anorthosite, a material made up chiefly of plagioclase and far richer in aluminium and calcium than the mare basalts. This prediction was almost correct. The major rock type found when Apollo 16 visited the Cayley highland formation was anorthositic gabbro, which contains more pyroxene than anorthosite and possibly was formed under conditions of slower cooling leading to larger and more evenly shaped crystals. All the mare regions of the Moon contain a proportion of highland materials which have been thrown there by major impacts. Consequently a second highland rock type known as "KREEP", rich in potassium

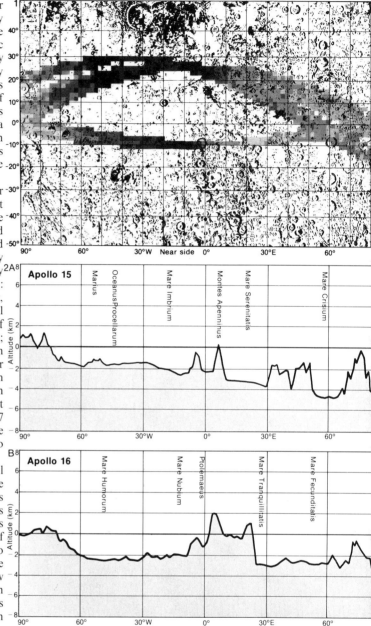

1. Radioactive regions
The orbiting command modules of Apollo 15 and 16 carried γ-ray spectrometers which measured the level of natural radioactivity in the lunar soil. The results indicate the concentration of thorium, uranium and potassium, since it is the decay of these elements that is the main source of the radioactivity. In this geochemical map the darkest shading represents the highest level of radioactivity, revealing the distribution of KREEP—which is rich in the elements concerned— on the lunar surface.

2. Altitude profiles
These graphs show the altitudes of features relative to a sphere of radius 1.738 km.

3. KREEP sample
KREEP is one of the two types of rock characteristic of the lunar highlands. This sample was brought back by Apollo 15.

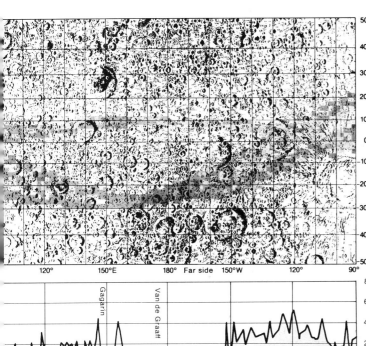

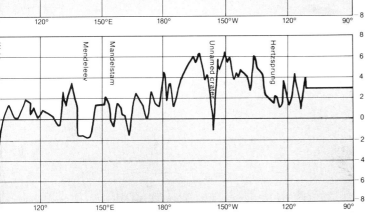

Gagarin, Van de Graaff, Mendeleev, Mandelstam, Unnamed crater, Hertzsprung

(K), rare earth elements (REE) and phosphorus (P), was soon recognized among the Apollo 12 rocks and soils. KREEP is concentrated among other places at Fra Mauro, the Apollo 14 site, and may represent material excavated from Mare Imbrium. The importance of KREEP is that it must have been formed by partial melting of the primitive crust—partial melting being a process that concentrates elements such as potassium with large atomic diameters. KREEP is also rich in radioactive elements uranium and thorium, therefore its distribution on the lunar surface may easily be mapped by orbiting geochemical instruments (*see* diagram 1).

Distribution

Altogether the highlands cover 85 percent of the total surface of the Moon, but they are not evenly distributed. The roughest part of the Earth-turned hemisphere is the south-east quadrant, where there are few walled formations with dark interiors (Stöfler, 145 km in diameter, is an exception). In the south-west highlands there are large, light-floored enclosures such as Bailly, almost 300 km across—more than half the diameter of the Mare Crisium—which has been aptly described as "a field of ruins". There are also highland areas around both the poles. The north polar crater has been appropriately named Peary, in honor of Robert Edwin Peary, the famous Arctic explorer.

The far side of the Moon is made up almost exclusively of terrae, with an average elevation of 5 km above the mean radius of the Moon's globe. Instead of maria, there are huge, circular depressions such as Apollo, Hertzsprung and Korolev, which are of marial size in many cases but are not filled with basalt. The regions furthest away from the regular ringed maria (Imbrium and Orientale in particular) have presumably been last disturbed by the events that produced the near-side basins and showered ejecta in all directions. Unfortunately there is as yet no chance of obtaining samples from them, since landing on the Moon's far side would be too hazardous to attempt in the present stage of technological development, although doubtless it will be done eventually. The far side of the Moon would be an ideal site for a radio astronomy observatory, since it would be shielded from the interference of all Earth transmissions.

The general aspect of the far-side uplands differs markedly from that of the terrae on the familiar hemisphere, as has been stressed by all the astronauts who have had direct views of it (the crews of Apollo 8, and 10 to 17 inclusive). It is not only the absence of maria that distinguishes it; there are differences in tone and in crater arrangement, and apparently the grid system is very much less marked. Of the crater-valleys, the most striking is that associated with the giant walled plain Schrödinger, at latitude 75°S.

One characteristic of all the uplands, on both the near and the far sides of the Moon, is the absence of mountain ranges of terrestrial type. On the Moon there is a clear distinction between highlands and mountains. The main lunar ranges form the boundaries of the circular maria and were produced in association with these phenomena. The highlands are the remnants of the lunar crust.

4. Highland terrain
The densely cratered region to the north-east of Tsiolkovsky on the far side of the Moon provides a good example of the appearance of the lunar terrae. The materials in this kind of region are thought to be the oldest on the Moon, but cratering has altered the rocks from their original forms.

4

Chemical composition of rock samples (weight%)	Lunar Highland Rocks		Average Earth Continental Granite
	Anorthositic gabbro	KREEP basalt	
SiO$_2$	44.5	48.0	72.3
TiO$_2$	0.39	2.1	0.3
Al$_2$O$_3$	26.0	17.6	14.0
FeO	5.77	10.9	2.4
MnO	0.07	0.14	0.05
MgO	8.05	8.70	0.5
CaO	14.9	10.70	1.4
Na$_2$O	0.25	0.70	3.1
K$_2$O	0.08	0.54	5.1
Cr$_2$O$_3$	0.06	0.18	—

Craters

Lunar craters have sunken floors, with rims rising only to modest heights above the outer surface. They are generally very shallow in relation to their diameters, so that it is quite wrong to think of them as resembling steep-sided mine-shafts. The vast structure Bailly, for instance, has a diameter of 295 km and a depth of 3.6 km, a ratio of 82 : 1; moreover, as with all lunar craters, the slopes of the walls are fairly gentle. Clavius has massive, regular walls, but the ratio of diameter to depth is only 47 : 1. There is, however, a tremendous variety in the forms, sizes and individual characteristics of walled formations. Some have central elevations, but these never reach the height of the ramparts; others, such as Timocharis and Lambert in the Mare Imbrium, have central craters; there are also concentric craters, of which Vitello and Taruntius are good examples. With some of the larger structures there are inner rings of peaks, rather different from the inner walls of the concentric craters. It is very rare to find a crater with both an inner peak-ring and a central elevation, although a few examples are known, notably Antoniadi (140 km in diameter) and Compton (175 km) on the Moon's far side.

Many other characteristics are shown by the craters. Some, such as Grimaldi and Plato, have dark, mare-type floors; there are the brilliant craters, notably Aristarchus, which is often visible when illuminated only by earthshine and has frequently been mistaken for a volcano in eruption. Then there are the craters that are centers of bright ray systems; Tycho and Copernicus are the most prominent, although there are many others. Tycho's rays do not issue directly from the center of the crater, but are mainly tangential to the walls.

Crater origins

The question of the origin of the large lunar craters has caused a great deal of discussion. It is now most generally accepted that meteorite impact is the predominant crater-forming mechanism. However, features also exist having morphologies that can only be interpreted as volcanic.

Crater origins may be recognized by the following general characteristics:

Impact	*Volcanic*
– asymmetric rim	– inverted V-shaped rim
– central peak	– no central peak
– interior terracing	– minimal terracing
– rough hummocky floor	– smooth lava-flow interior
– extensive irregular ejecta blanket	– generally smooth tapering exterior, sometimes channelled
– clusters of associated secondary craters, often aligned	– no apparent related features

The basically circular shape of most lunar craters is no bar to the impact theory. When a missile strikes, its kinetic energy is transformed into heat, making the missile equivalent to a powerful explosive, and a circular crater is the result, even if the missile comes in at an angle. Central peaks are attributed to rebound. The ratios of depth to diameter of lunar craters match those of terrestrial explosion features. The juxtaposed craters Aristarchus and Herodotus (*see* diagram 3) may be representative of the two genetic types, the former being produced by impact.

With recourse to "ground truth" measurements from sample analysis and study, the recognition of impact features becomes much easier. Evidence of the impacting body may be found by chemical study of the resulting ejecta. The existence of shock-lithified breccias within the ejecta blanket also provides evidence of an impact origin.

Among terrestrial impact craters, the most famous is that in Arizona. Its diameter is 1,265 m, and the crater was certainly produced by the impact of a meteorite about 22 thousand years ago. Another unquestioned meteorite crater is Wolf Creek in Australia.

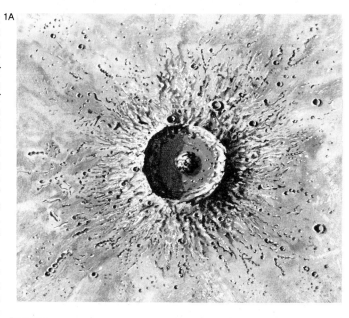

1. Crater morphology
Walled formations on the Moon are of various types, although in a typical example the basic form is circular (**A**). The main rampart may be very massive and terraced; the floor is sunken, and may contain a central peak or group of peaks. The surrounding area is generally covered by a blanket of ejecta, and pitted with secondary craters. Seen in profile (**B**) a crater is shallow, although smaller craters are comparatively deep relative to their diameters.

2. Impact cratering
Craters remitting from the impact of objects with velocities greater than 15 km s⁻¹ must be common on the Moon. Laboratory impact studies using a hyper-velocity gun have reproduced features strikingly similar to some of those on the Moon, although on a much smaller scale. The formation resulting from such impacts is circular even when the projectile strikes the surface at an oblique angle. The missile strikes the surface (**A**) sending shock-waves into the target and the projectile itself. The resulting stresses cause the projectile to be vapourized, and material is ejected at very high velocities from around the projectile. At the next stage material is ejected at lower velocities, excavating the crater (**B**) as a plume of material is sent outwards from the center (**C**). Finally the ejected material falls back to the surface and the rim of the crater is formed by uplift (**D**). Secondary craters are formed by material falling back, and it is assumed that the central peaks are produced by a kind of rebound effect.

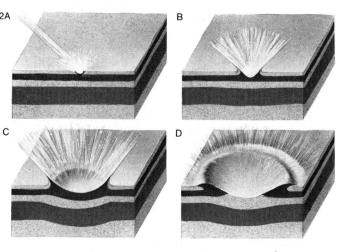

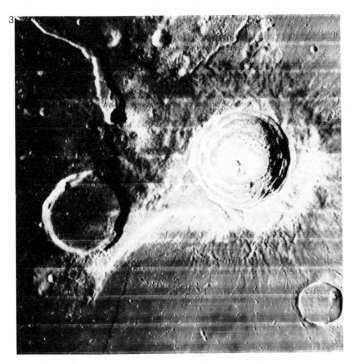

The absence of a substantial number of undisputed impact craters on the Earth to compare with those of the Moon is generally accounted for by the effects of erosion due to the terrestrial weather and other geological influences. The Moon, lacking any appreciable atmosphere, has retained the evidence of meteoritic bombardment. In the present era, moreover, the Earth's atmosphere protects it from the impacts of small meteorites, which are burnt up before reaching the surface. However, care must be taken not to jump to conclusions, the nature of many terrestrial features is also hotly disputed.

Returning to the Moon, the impact theory is not without problems. For example, summit pits are often found on central peaks, and it seems that these are too numerous and too symmetrically placed to be due to chance impacts. Moreover, if impact were the sole cratering agency, then the distribution of the larger formations would be expected to be random, but this is not the case. When one crater breaks into another it is almost always the smaller formation which is the intruder, so that presumably it is also more recent. The rule applies in almost all cases; a striking example is Thebit, 60 km in diameter, which is broken by a smaller crater (Thebit A), which is in turn interrupted by a still smaller structure (Thebit F). On the other hand, this phenomenon could reflect a decay in the size distribution of impacting bodies.

Lines of vast enclosures also occur, such as that running down the Moon's eastern limb as seen from Earth (Furnerius, Vendelinus, Petavius, Langrenus and through to Mare Crisium and Cleomedes). These lines appear to be related to the central meridian of the near hemisphere, which can hardly be coincidence; it has been attributed to the effects of the Earth's pull, and certainly there is every reason to assume that the lunar rotation has been synchronous since very early in the Moon's evolution.

An attempt was made by the last Apollo mission to visit an area that was thought to be a likely candidate for recent volcanic activity. However, the dark mantling material at the Taurus–Littrow site turned out to be essentially due to the abundance of glasses very rich in titanium rather than fresh lava flows. The most common features of volcanic activity on the Moon must be the "wrinkle-ridges", which abound on the mare flows, and the rills and crater chains, which are presumably associated with lava tubes. ("Lava tubes" arise from the formation of a crust of congealed lava over a channel of flowing lava; the channel may subsequently empty of fluid lava, leaving a hollow tube whose thin roof may later collapse.) Most of these features are very old indeed. With magmas of low viscosity the building up of mountains of lava and ash into graceful forms such as Vesuvius is impossible. Volcanic features associated with the formation of the crust are essentially absent from the highlands, although that is not to say that intense bombardment has not erased all trace of pre-existing evidence of activity.

Crater counts
On the assumption that the impact of meteorites is the most important mechanism, it is possible to gain a great deal of information from counting small craters on various features: based on theoretical models of what the meteorite influx should have been, the ages of different areas may be assessed. The method certainly works very well at the microscale when applied to lunar rocks, but there are differences in applying it to the Moon's surface. One difficulty is that some of the craters included in the count may be of volcanic rather than impact origin; another is that a surface eventually becomes saturated with craters, at which point fresh impacts effectively erase earlier craters as quickly as they create new ones. Consequently there is no apparent net change in the number of craters of a given size. However, crater counting, fortified by the knowledge gained from radiometric dating of lunar materials, is currently the only available method of estimating the age of surface features on other planetary bodies.

3. Herodotus and Aristarchus
These two adjacent craters are often cited as typical examples of formation by, respectively, volcanic and impact mechanisms. The inside and outside of the rim of Herodotus are approximately symmetrical; the crater has no central peak and no secondary craters appear to be associated with it; there is very little terracing, and the floor, filled with lava, is comparatively smooth. Aristarchus, on the other hand, has the asymmetric rim typical of an impact crater, with terraced inner walls and a thick blanket of ejecta material on the outside, as well as numerous secondary craters. Aristarchus also has a central peak. The large rill in the photograph is Schröter's Valley.

4. Terrestrial craters
Various craters on Earth have been identified as being of impact origin: two proven examples, where remnants of the meteorite that struck have been found on the site, are the Meteor Crater in Arizona (**A**), and Wolf Creek in Australia (**B**). Meteor Crater is 1,265 m in diameter and 174 m deep; its estimated age is 22,000 years. Wolf Creek is 853 m in diameter and 46 m deep. Volcanic craters (apart from those on volcanic cones) are also found on Earth; a good example is Hverfjall in Iceland (**C**). Craters have also been produced artificially by the use of explosives (**D**): this example, in western Canada, bears a strong likeness to the far-side crater Tsiolkovsky (*see* page 81), although it is very much smaller in scale (about 100 m in diameter and 6.5 m deep).

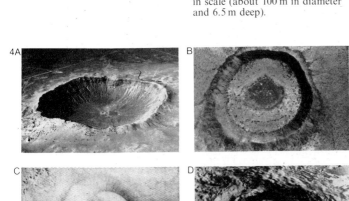

Unusual Features

There are various features on the Moon's surface that do not readily fit into any of the broad categories of feature types, or which have some peculiarity that distinguishes them from the norm. There are, for example, several craters or depressions with highly irregular shapes, suggesting that they were formed in different ways from the majority of craters; many of them may be volcanic in origin. However, any suggestions about the formation of these features must remain speculative. A selection of photographs showing some of the most interesting unusual features is provided on this page.

1. Aratus CA
Comparable depressions are often found at the head of lunar rills, although none of the others have such complicated shapes as this example, which is about 6 km in length. Features of this kind are probably formed by the collapse of lava tubes.

2. Irregular crater pair
This pair of unusually shaped craters is thought not to have been formed directly by the impact of bodies from space, although it may be the result of secondary impacts. It is located near Guericke, and has a total length of approximately 18 km.

3. D-shaped depression
At least three different types of material are apparent in this strange feature, which seems to be unique. The depression is located to the west of the Mare Serenitatis, at the foothills of the Haemus mountains. It is about 3 km in width, and is thought to be volcanic.

4. Flooded crater
This 4 km feature, situated in the Mare Serenitatis, appears to be the remains of a crater that has been filled with molten lava after its formation on a preexisting surface. For the structure to have survived, a certain amount of subsidence has subsequently taken place, perhaps as loosely packed regolith was compacted.

5. Irregular crater
This crater, approximately 10 km by 16 km, is located on the floor of Barbier on the far side of the Moon. The peculiar, almost rectangular shape at one end has not been accounted for, but may be related to faults in the lunar crust. There is evidence of ejecta material.

6. Mare units
This photograph shows a partially filled crater lying on the border of two different mare units; the units meet along a line running roughly parallel to the straight rill which touches the edge of the crater. The lighter unit is the younger of the two and has filled in much of the crater. The crater itself has a diameter of approximately 9 km.

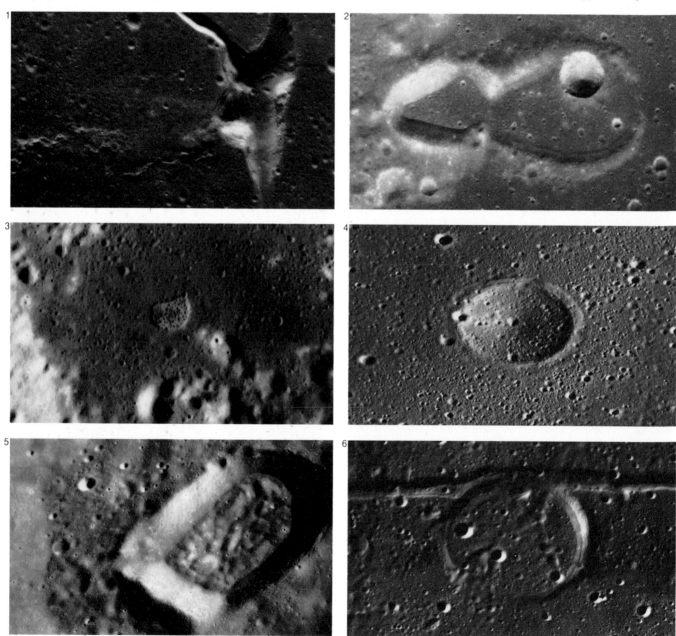

Color Plates

Mare Crisium and Mare Fecunditatis
Apollo 8
December 1968

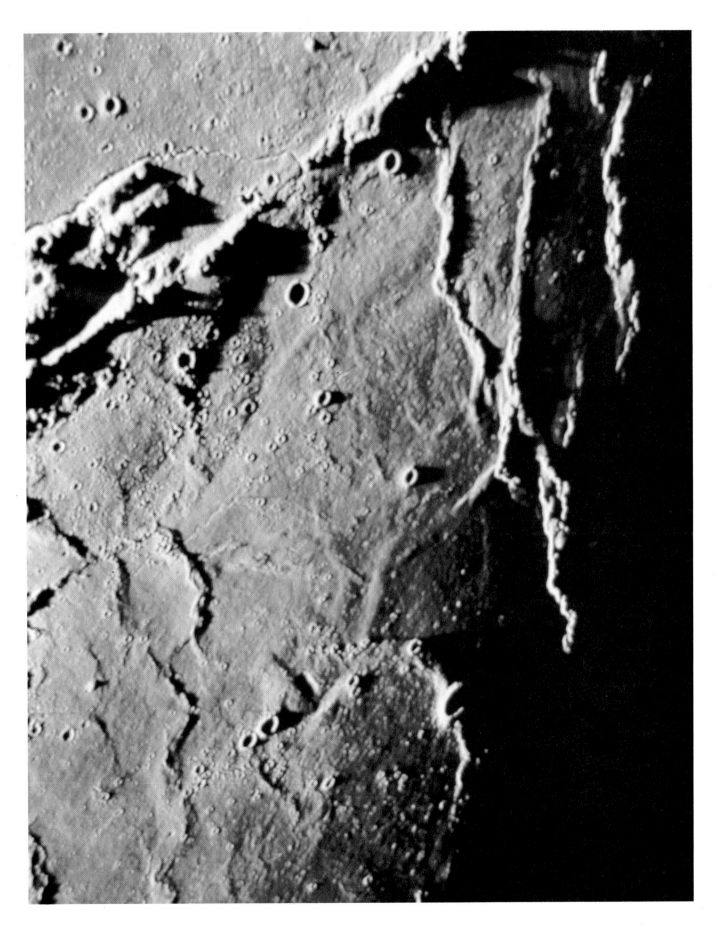

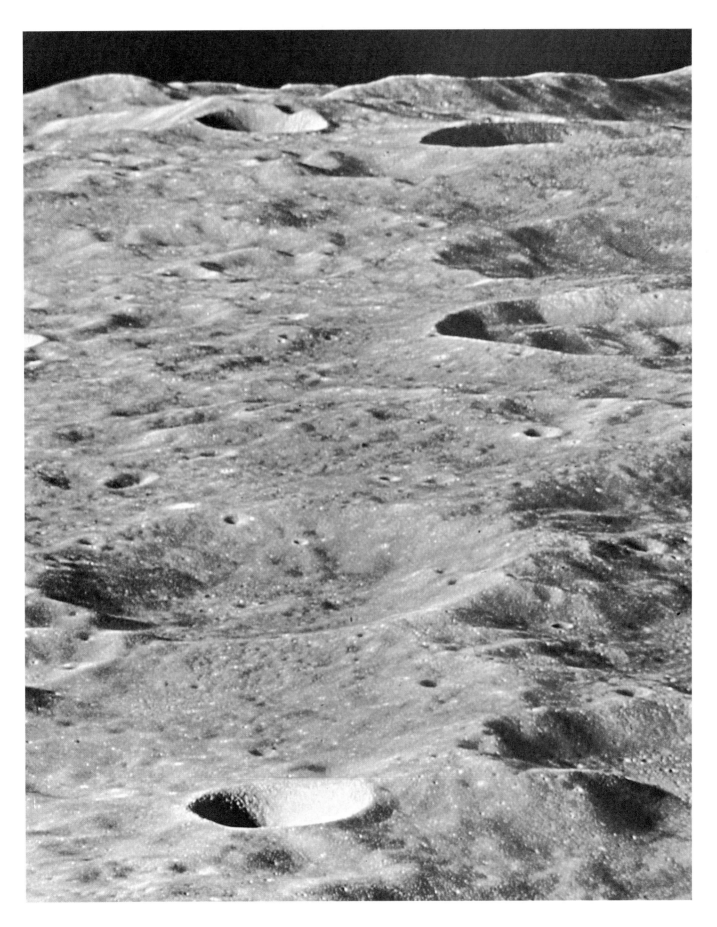

Copernicus (top)
Apollo 17

Van de Graaff (bottom)
Apollo 17

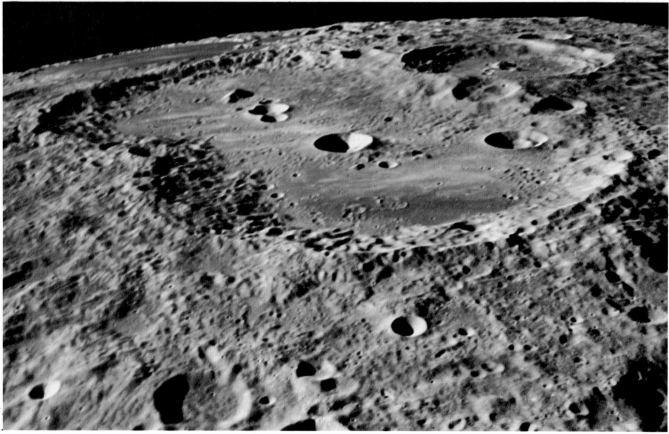

Eratosthenes (top)
Apollo 17

Doppler (bottom)
Apollo 17

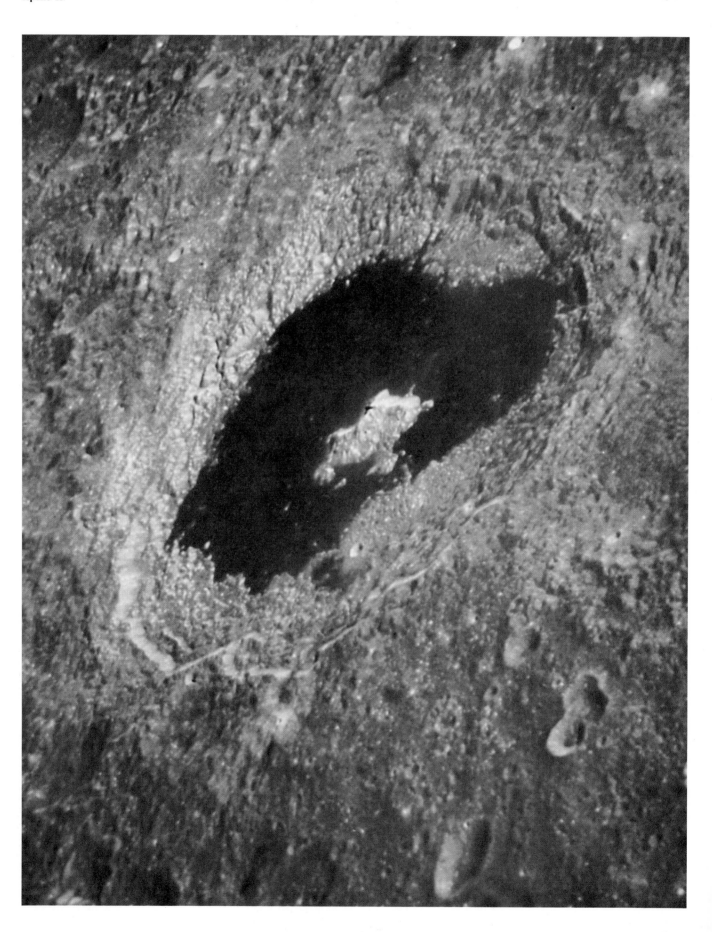

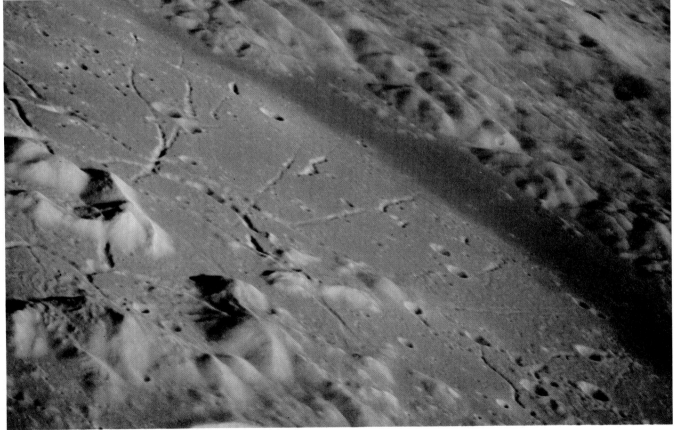

Humboldt (top)
Apollo 15

Detail of Humboldt (bottom)
Apollo 15

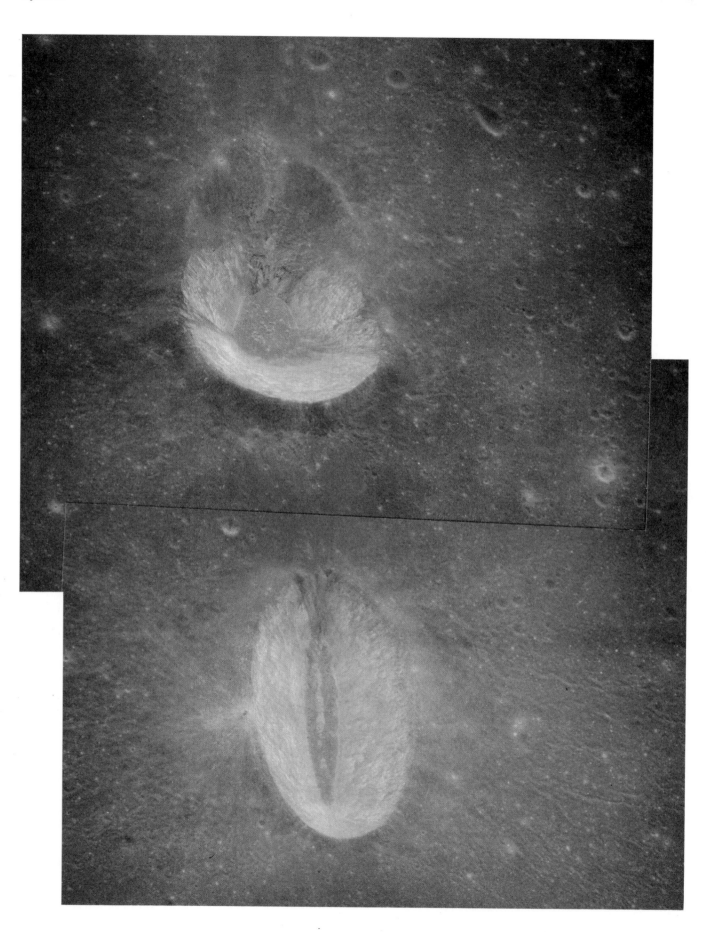

Taurus-Littrow (detail)
Apollo 17

Lunar basalt
Apollo 11
Reflected light (top), polarized light (bottom)

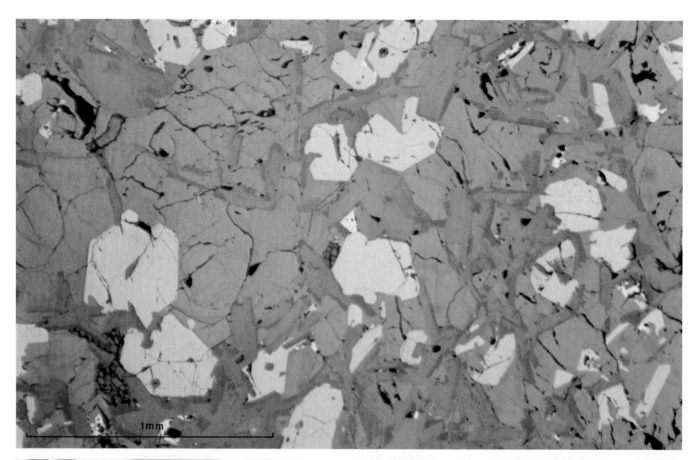

1mm

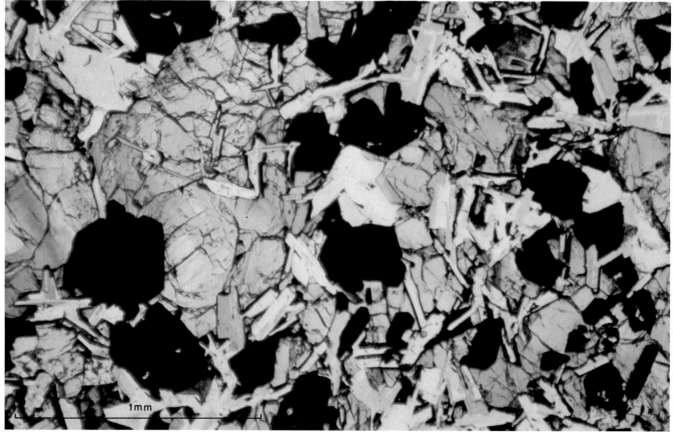

1mm

Notes to Color Plates

Page 33. Taken from the Lunar Module of Apollo 11 during the return journey at a distance of 16,000 km. The disc is centered at about 7° E.

Page 34. Detail of north-east section of the near side, showing two prominent maria, Crisium and Fecunditatis.

Page 35. Mare region to the north-west of Aristarchus, photographed from the Command Module of Apollo 15.

Page 36. Vertical view of highland terrain on the far side of the Moon (3° S, 155° E) photographed by Apollo 8.

Page 37. Oblique view of highland terrain; the region shown is close to that in the previous photograph (10° S, 155° E).

Page 38 (Top). Oblique view of the crater Copernicus (10° N, 20° W) showing central peaks and terraced walls. The small crater to the right of the foreground is Gay-Lussac A, while the slightly larger crater in the bottom right-hand corner is Gay-Lussac. (Bottom). The large, irregularly shaped feature is Van de Graaff (27° S, 172° E); several smaller craters intrude, the largest (towards the top right-hand corner in the photograph) is Birkeland (30° S, 174° E).

Page 39 (Top). Eratosthenes (15° N, 11° W). The large crater on the limb in this photograph is Copernicus. (Bottom). The main feature, Doppler (13° S, 160° W), adjoins a much larger crater, Korolev, which can be seen stretching away towards the limb. The walls of Korolev are comparatively worn down, and both Doppler and Korolev are pitted with many smaller craters.

Page 40. Tsiolkovsky (21° S, 128° E) is one of the most prominent and unusual features of the far side of the Moon.

Page 41 (Top). Humboldt (27° S, 81° E) has a remarkable system of radial and concentric fractures on its floor. (Bottom). Detail of Humboldt, showing the fractures.

Page 42. The twin craters Messier (above) and Messier A (lower) are situated in the Mare Fecunditatis (at 2° S, 48° E). Messier A has a distinctive pattern of rays.

Page 43 (Top). Parry (8° S, 16° W) is the large walled plain in the foreground of the photograph; Fra Mauro is to the right, and Bonpland can be seen in the background. A graben-like rill, Rima Parry, runs across the picture, hitting a small, well-formed crater on the common wall of Parry and Bonpland. (Bottom). This unnamed crater on the far side of the Moon was photographed from the Lunar Module of Apollo 10 during its descent to within 15 km of the lunar surface.

Pages 44 and 45. Schröter's Valley. The crater at one end of Schröter's Valley is known as the Cobra Head; such irregular depressions are characteristic of sinuous rills. Schröter's Valley is one of the largest sinuous rills on the Moon.

Page 46. Vertical view of Taurus-Littrow (20° N, 31° E), the site of the Apollo 17 landing (indicated by a white arrow). The mountain mass to the bottom right of the photograph is called South Massif.

Page 47. View from the lunar surface of the valley of Taurus-Littrow, looking north-east. In the foreground is a large boulder field.

Page 48 (Top). Mare basalt from Apollo 11 mission, taken in reflected light (sample number 10003,152). The following minerals are easily identifiable: plagioclase (dark grey, elongate); pyroxene (light grey, surrounding plagioclase); ilmenite (light blue); cristobalite (SiO_2) (very dark grey-brown, with many cracks—one grain may be seen near the center, several others in the lower left-hand corner); troilite (FeS) (bright pinkish mineral, uncommon in this sample—one small grain near the center). Other minerals are discussed in the main text—*see* pages 22–23. (Bottom). Same sample of mare basalt as in the previous photograph, this time viewed under plane polarized light. Plagioclase shows up as white, pyroxene as multicolored, ilmenite as black, cristobalite as white with cracks, and troilite as black.

Cartography

Between 1966 and 1967 the five Lunar Orbiters obtained photographic coverage of virtually the entire surface of the Moon. The Orbiter program was, in fact, specifically designed to provide the photographic data necessary to map the Moon, paying particular attention to potential landing sites for later Apollo missions. The first three Orbiter spacecraft were placed in near-equatorial orbits, while the remaining two followed orbits that took them over the poles. The photographs were made on strips of 77 mm film and were processed automatically onboard the spacecraft. These were then electronically scanned in small sections and transmitted back to Earth, and the images they provided form the basis of subsequent lunar cartography.

In spite of the great success of the Lunar Orbiter Mission the photographic system had a few drawbacks, and it was difficult to make accurate measurements of the positions of lunar features from Orbiter photographs. Moreover, relatively little stereoscopic coverage was obtained. The systems designed for the Apollo mission had the advantage of being able to return photographic films to Earth for processing, and employed a range of camera types to improve on and supplement the earlier photography. The orbiting Apollo modules photographed nearly 20 percent of the lunar surface in detail.

The most sophisticated mapping equipment was carried by the last three missions, Apollos 15, 16 and 17. These were equipped with two high-resolution cameras specifically designed to provide photographs suitable for cartographic purposes: a mapping camera system and a panoramic camera. The panoramic camera provided stereographic images of relatively large areas down to a resolution of 1 or 2 m. The mapping camera also made stereoscopic images (of slightly lower resolution) as well as providing the data necessary to reconstruct with greater accuracy the geometry of the lunar surface. To achieve this, the system's terrain camera was coupled to a stellar camera and a laser altimeter to record the orientation and altitude of the spacecraft at the time of each exposure (*see* diagram 2).

Map projections

Three types of map projections are used in this atlas. The Moon's near-side hemisphere is shown first in an orthographic projection (*see* diagram 3) which depicts the Moon as it appears from Earth, and which is therefore the most convenient type of map for the telescopic observer. Detailed views of the same map occupy pages 54–57, 60–63 and 66–69, and each group of maps is followed by a two-page spread of telescopic views of the region photographed from Earth. The perspective of the photographs thus corresponds to that of the preceding maps.

However, because features near the edge of an orthographic projection appear foreshortened, these maps do not give a good view of features near the Moon's limb, where circular craters, for example, appear as ellipses. To give a better impression of how the feature would look from directly overhead, a pair of conventional Mercator projections is used, one showing the near side (page 72–73) and the other showing the far side (page 78–79). Each Mercator map is followed by four pages of photographs taken by one of the lunar space missions, so that, once again, the photographs match the maps. Mercator projections, however, although relatively free of distortion in the equatorial regions, produce distortions in the higher latitudes and are unsuitable for showing the poles. These regions are therefore presented in polar stereographic projections (pages 84–85), which overlap the high latitudes of the Mercator maps by a few degrees.

In reality, the Sun's rays strike different parts of the globe at different angles, varying continuously with time. Some features are therefore easier to observe at some times than at others. In order to show each feature as clearly as possible, the lighting for these maps has artificially been treated as though the Sun was at a constant, low angle across the entire surface of the Moon.

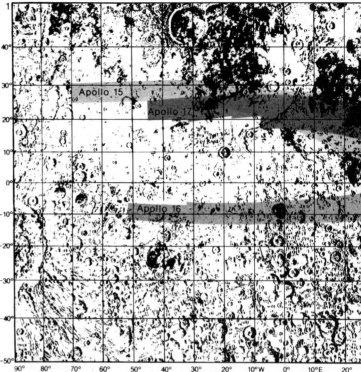

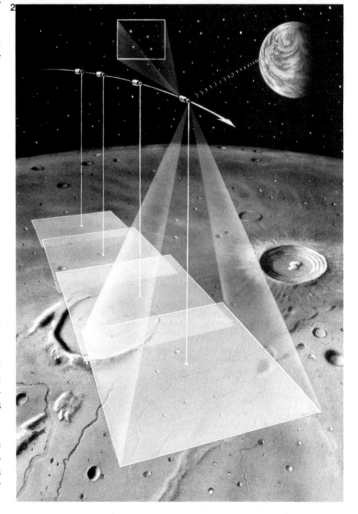

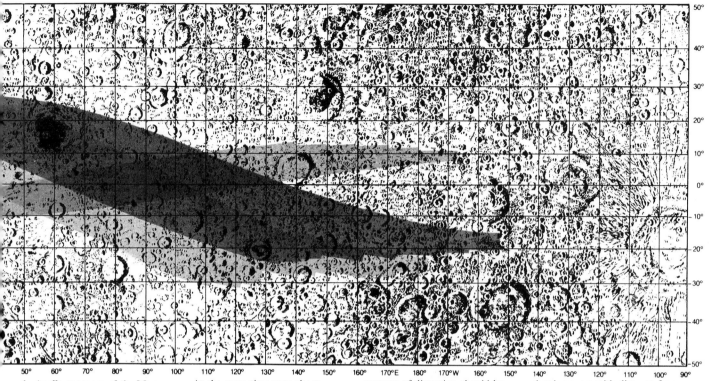

1. Apollo coverage of the Moon
The regions of the lunar surface photographed by the mapping cameras of Apollo missions 15, 16 and 17 are shown as shaded areas. Only the regions photographed in sunlight from directly overhead are plotted.

2. Apollo mapping camera system
The system comprised three essential components: a stellar camera, a laser altimeter and the mapping camera itself. The mapping camera recorded sequences of overlapping images which formed a continuous strip.

At the same time very short pulses of ruby laser were emitted parallel to the camera's line of sight; the time taken for the laser to be reflected back to the altimeter gave an accurate measurement of the spacecraft's altitude above the lunar surface, while its orientation could be determined from images of the stellar background made by the stellar camera.

3. Cartographic projections
When features on a spherical surface are represented on a plane it is inevitable that a certain

amount of distortion should be introduced. Each type of map projection employs a different mathematical construction to preserve the most important spatial relationships, according to the map's intended purpose. The orthographic projection (**A**) represents the globe as a flattened disc; the distortion is the same as that seen by an observer looking at the Moon from Earth, with craters near the limb being flattened into ellipses. The Mercator projection (**B**) is a special type of cylindrical projection. It has a latitude scale

that increases with distance from the equator, calculated in such a way that any straight line represents a constant bearing. To achieve this unique property, the scales have to be constructed mathematically. Polar stereographic projections (**C**) show the globe projected from one of the poles into a plane tangential to the opposite pole. In this atlas they are used to supplement the Mercator projections, showing the north and south polar regions that would appear excessively distorted on a Mercator map.

3A   B   C

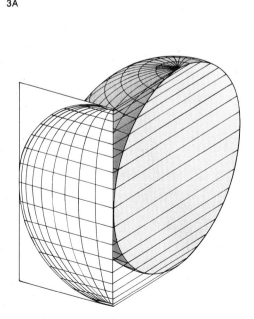

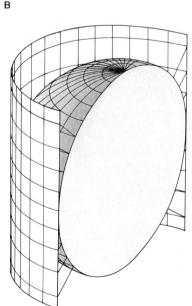

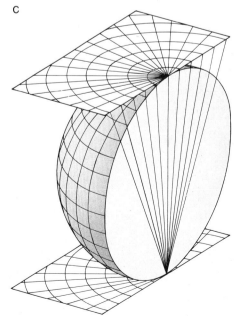

Near Side Orthographic

North-west
This section is dominated by the
Mare Imbrium, the largest and
most important of the Moon's
regular "seas". Its borders include
the Apennines, the Alps, the Jura
Mountains and the Carpathians;
the Apennines are the most
impressive of all the lunar ranges.
Leading out of the Mare Imbrium
is the beautiful Sinus Iridum.
Other mare areas include the
Sinus Roris, part of the Mare
Frigoris and a small part of the
Oceanus Procellarum. Walled
formations include the dark-
floored Plato, and Archimedes on
the Mare Imbrium (the other
formations of the Archimedes
group, Aristillus and Autolycus,
are just in the north-east section
of the map). Of special
importance is Aristarchus, the
brightest point on the entire
Moon; nearby is the darker-
floored Herodotus, together with
the magnificent Schröter's Valley.

Center west
Much of the huge Oceanus
Procellarum is to be found in this
section, with part of the Mare
Nubium and smaller marial areas
such as the Sinus Aestuum. The
whole region is dominated by
Copernicus, the magnificent ray-
crater with its high, terraced walls
and central mountains; Kepler is
another ray-crater, and there is
yet another, Olbers, closer to the
limb. Other very important walled
formations are: Ptolemaeus,
Alphonsus and Arzachel, which
form a splendid chain; Grimaldi,
Riccioli and Gassendi on the
northern border of the Mare
Humorum. Part of the Apennines
range extends into the section,
ending near the deep, well-marked
formation Eratosthenes. The
Riphaean Mountains are quite
prominent, as are the Carpathian
Mountains forming part of the
border of the Mare Imbrium.

South-west
The main "sea" areas here are
those of the Mare Humorum and
part of the Mare Nubium; on the
Mare Nubium is the celebrated
Rupes Recta or "Straight Wall",
the most famous fault on the
Moon. There are few craters on
the Mare Humorum (Gassendi, at
the northern edge, is just in the
center section of the map), but
there are some well-marked
"bays", such as Doppelmayer and
Hippalus. High peaks, associated
with the Mare Orientale, lie along
the limb. The southern area is
mainly highland, with some huge
walled plains, such as Schickard,
Clavius and the irregular Schiller,
as well as the great chain
composed of Purbach,
Regiomontanus and Walter; but
the most prominent feature in this
region is the ray-crater Tycho.
Seen under high light, the Tycho
rays dominate the entire area, and
indeed much of the Moon's disc.

North-east

The main "sea" area in this section is the Mare Serenitatis, almost all of which is included. There are few major formations on its floor; the most prominent is the small but easily recognized Bessel, and here too is Linné, the minor feature once suspected of having shown definite change in the nineteenth century, although this is now discounted. A small part of the Mare Crisium is included, together with a large section of the irregular Mare Frigoris; on the limb is the well-marked Mare Humboldtianum. Of the walled formations, Posidonius on the edge of the Mare Serenitatis is prominent; so are the Atlas–Hercules, Aristoteles–Eudoxus, and Aristillus–Autolycus pairs. The Alpine Valley comes into this section, and there is an important system of rills associated with the well-known crater Bürg.

Center east

There are wide marial areas here. Almost all of the well-marked Mare Crisium is included, and the whole of the Mare Tranquillitatis, together with much of the Mare Fecunditatis and the Mare Nectaris, and the smaller Mare Vaporum. Objects of special interest include the Messier twins on the Mare Fecunditatis, the ray-crater Proclus near the distinctive Palus Somnii, and the prominent rills associated with Hyginus and Ariadaeus. The great walled formations Langrenus and Vendelinus lie on the boundary of the Mare Fecunditatis. Close by the border of the Mare Nectaris are the prominent craters Theophilus, Cyrillus and Catharina, making up one of the most magnificent "chains" on the Moon. "Tranquillity Base", the landing site of Apollo 11, falls in this section.

South-east

This is the main upland region of the visible hemisphere of the Moon; there are almost no "sea" areas apart from the patchy, irregular Mare Australe near the limb. The Altai Scarp, associated with the Mare Nectaris system, runs in from the center section of the map, and there is also the crater-valley of Rheita, with the almost equally prominent valley associated with Reichenbach. Large walled formations include Petavius, with its magnificent internal rill; Furnerius and the huge, ruined Janssen. Right on the limb is Wilhelm Humboldt, always hard to study from Earth because of its extreme foreshortening, but clearly shown on Orbiter and Apollo pictures; the floor contains many rills. Other major formations are Stöfler and Maurolycus. The whole area is crowded with walled formations of all types.

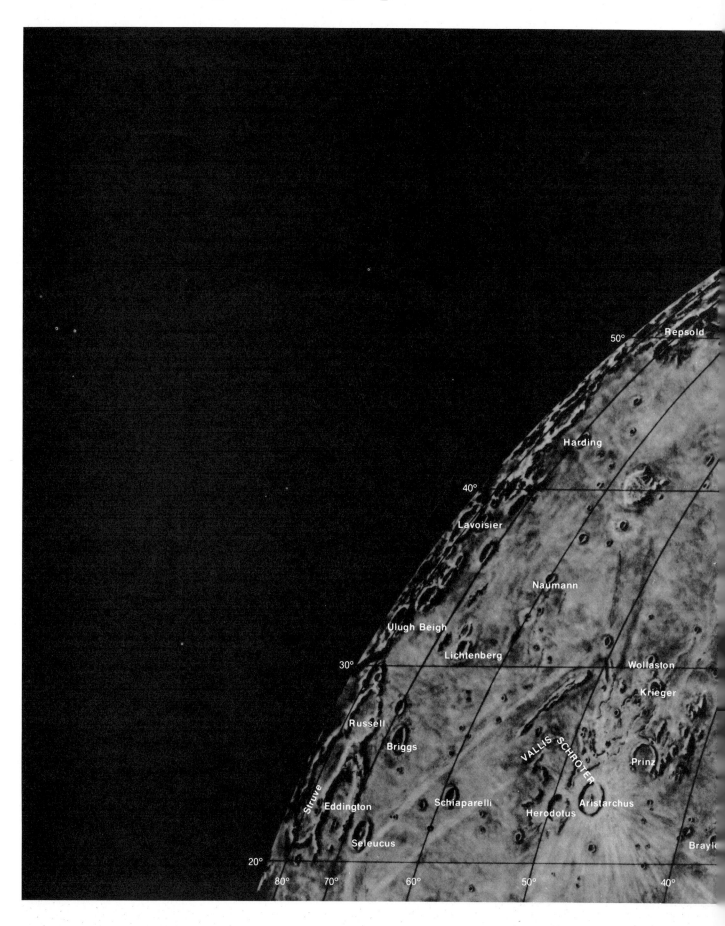

Repsold

50°

Harding

40°

Lavoisier

Naumann

Ulugh Beigh

Lichtenberg

Wollaston

30°

Krieger

Russell

VALLIS SCHRÖTER

Briggs

Prinz

Struve

Eddington

Schiaparelli

Aristarchus

Herodotus

Seleucus

Brayl

20°

80°   70°   60°   50°   40°

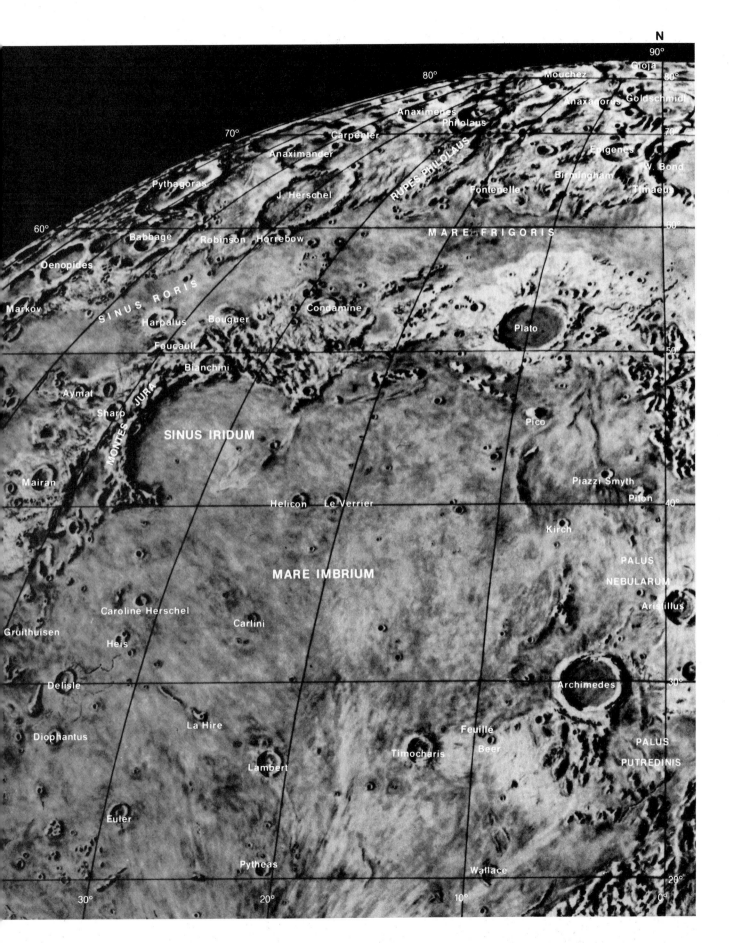

N

90°

80°

Gioja

Mouchez

Anaximenes

Carpenter

Philolaus

Anaxagoras

Goldschmidt

70°

Anaximander

J. Herschel

RUPES PHILOLAUS

Fontenelle

Epigenes

Birmingham

W. Bond

Pythagoras

Timaeus

60°

MARE FRIGORIS

Babbage

Robinson

Horrebow

Oenopides

SINUS RORIS

Markov

Condamine

Harpalus

Bouguer

Plato

Foucault

Bianchini

50°

Aymal

MONTES JURA

Sharp

SINUS IRIDUM

Pico

Mairan

Piazzi Smyth

Helicon

Le Verrier

Piton

40°

Kirch

PALUS

MARE IMBRIUM

NEBULARUM

Caroline Herschel

Aristillus

Carlini

Gruithuisen

Heis

Delisle

Archimedes

30°

La Hire

Diophantus

Feuillé

PALUS

Timocharis

Beer

Euler

Lambert

PUTREDINIS

Pytheas

Wallace

20°

30°

20°

10°

0°

55

N

90°
Hoja
80°   De Sitter   80°
                        Euctemon
                        Baillaud
Goldschmidt
                Barrow   Meton
70°   70°
                        Nelson   Arnold
W. Bond   Moigno
                                Kane
Timaeus   Democritus   Strabo   Belkovich
                                                        Thales
60°   60°
        Archytas   Sheepshanks   Gärtner   De la Rue   MARE HUMB
        Protagoras
                            Galle   Endymion

                        Aristoteles
50°   Mitchell   Baily
        VALLIS ALPES   Egede
                                            LACUS   Hercules   Atlas
                                        Bürg
                        Eudoxus   MORTIS
                                                                        Oerst
                            Lamech   Plana Mason   Williams
Piton   Grove   Ceph
40°   Cassini
                Calippus   LACUS SOMNIORUM
                                                                        Maury
PALUS   Daniell
NEBULARUM   Hall
    Aristillus   G. Bond
                                        Luther
                                                    Posidonius
        Autolycus   Kirchh
30°   Chacornac

                                    MARE SERENITATIS
PALUS   Linné   Le Monnier
PUTREDINIS   Röme
                                                    Bessel
            Aratus   Deseilligny   Littrow
    Conon
20°   0°   Sulpicius Gallus   20°   30°   Maraldi
            10°

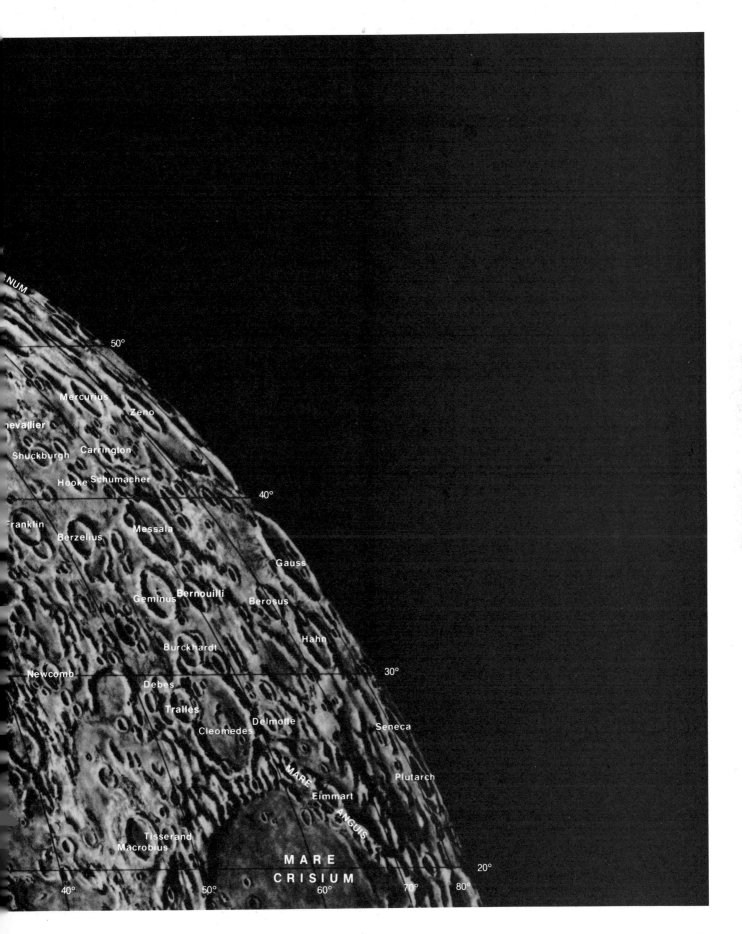

NUM

50°

Mercurius

Zeno

hevallier

Shuckburgh   Carrington

Hooke Schumacher   40°

Franklin

Berzelius   Messala

Gauss

Bernouilli

Geminus   Berosus

Burckhardt   Hahn

Newcomb   30°

Debes

Tralles   Delmotte

Cleomedes   Seneca

MARE   Plutarch

Eimmart

ANGUIS

Tisserand
Macrobius   MARE   20°
                            CRISIUM
40°   50°   60°   70°   80°

North Region

MONTES RECTI

SINUS IRIDIUM

Helicon   Leverrier

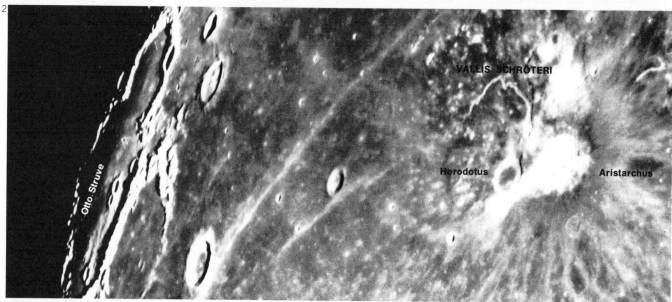

VALLIS SCHRÖTERI

Otto. Struve

Herodotus   Aristarchus

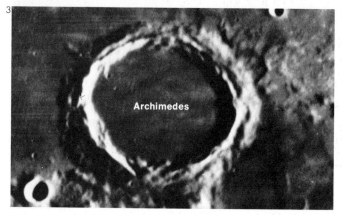

Archimedes

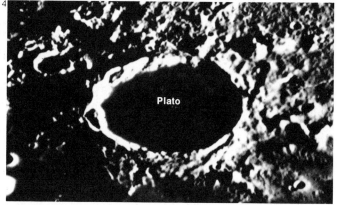

Plato

North-west region (left)

1. Key features: Sinus Iridum; Montes Recti; Helicon (diameter 29 km); and Leverrier (25 km). The Sinus Iridum is one of the most beautiful formations on the Moon, and under suitable conditions of lighting the Jura Mountains on its border seem to stand out from the terminator like a "jewelled handle".

2. Key features: Otto Struve (two rings, each about 160 km in diameter); Aristarchus (37 km); Herodotus (37 km); and Schröter's Valley. Aristarchus is so brilliant that it is often visible in earthlight; many TLP have been reported in and near it. The winding Vallis Schröteri is the best example of this type of feature on the Moon.

3. Archimedes (diameter 75 km). A splendid, circular formation on the Mare Imbrium, with rather low walls and a mare-type floor.

4. Plato (diameter 97 km). A regular walled plain, with a relatively smooth floor; one of the darkest areas found on the Moon.

North-east region (right)

5. Aristoteles (diameter 97 km); a great plain with walls rising to 3,350 m. The floor contains low hills; closely outside the east wall is Mitchell (diameter 19 km). Aristoteles makes a grand pair with its slightly smaller neighbour Eudoxus.

6. Key features: Bessel (diameter 19 km) and Linné (11 km). These lie on the comparatively level Mare Serenitatis. Bessel lies in the path of a long bright ray running across the Mare.

7. Bürg (diameter 48 km). Bürg has a concave floor, with walls rising to 1,800 m: it has a very large central mountain, which is crowned by a summit craterlet. Bürg lies inside a large ruined formation and is associated with a prominent system of rills.

8. Cleomedes (diameter 126 km): a magnificent enclosure near the border of the Mare Crisium. The walls average at least 2,700 m in height and are interrupted by a very deep crater, Tralles, which is 48 km in diameter.

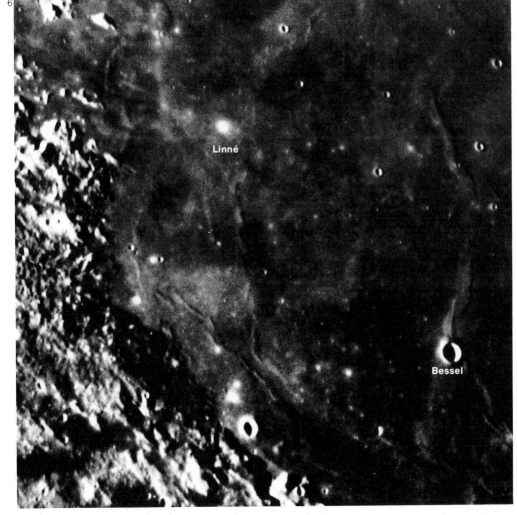

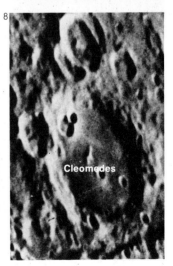

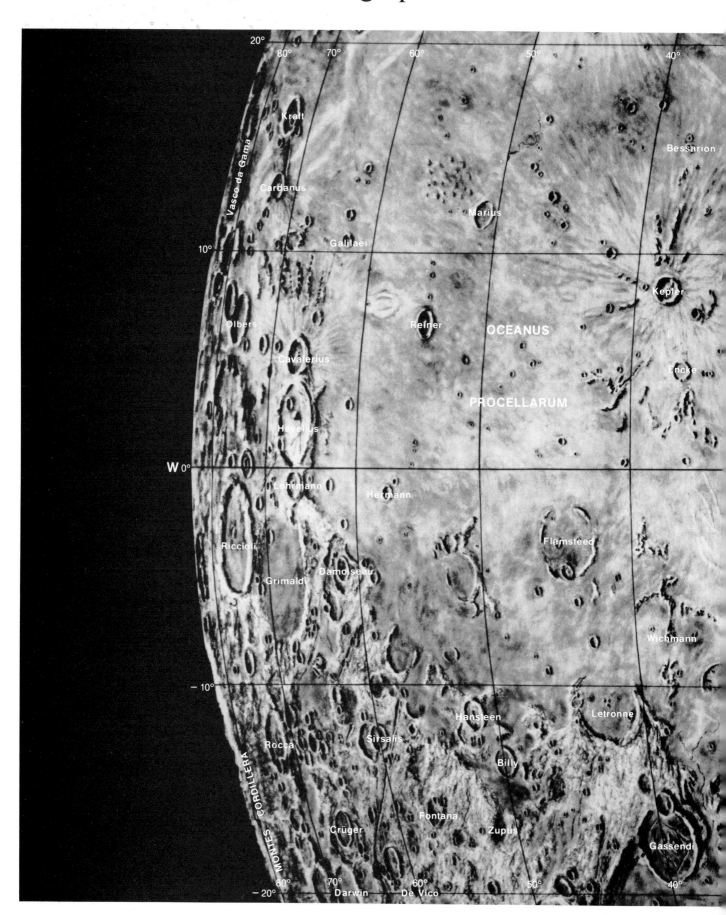

20°
80°
70°
60°
50°
40°

Kraft

Vasco da Gama

Cardanus

Bessarion

Marius

10°
Galilaei

Kepler

Olbers

Reiner

OCEANUS

Cavalerius

Encke

PROCELLARUM

Hevelius

W 0°

Lohrmann

Hermann

Riccioli

Flamsteed

Damoiseau

Grimaldi

Wichmann

− 10°

Letronne

Hansteen

Rocca

Sirsalis

Billy

CORDILLERA

Fontana

Zupus

MONTES

Crüger

Gassendi

80°
70°
60°
50°
40°

− 20°
Darwin
De Vico

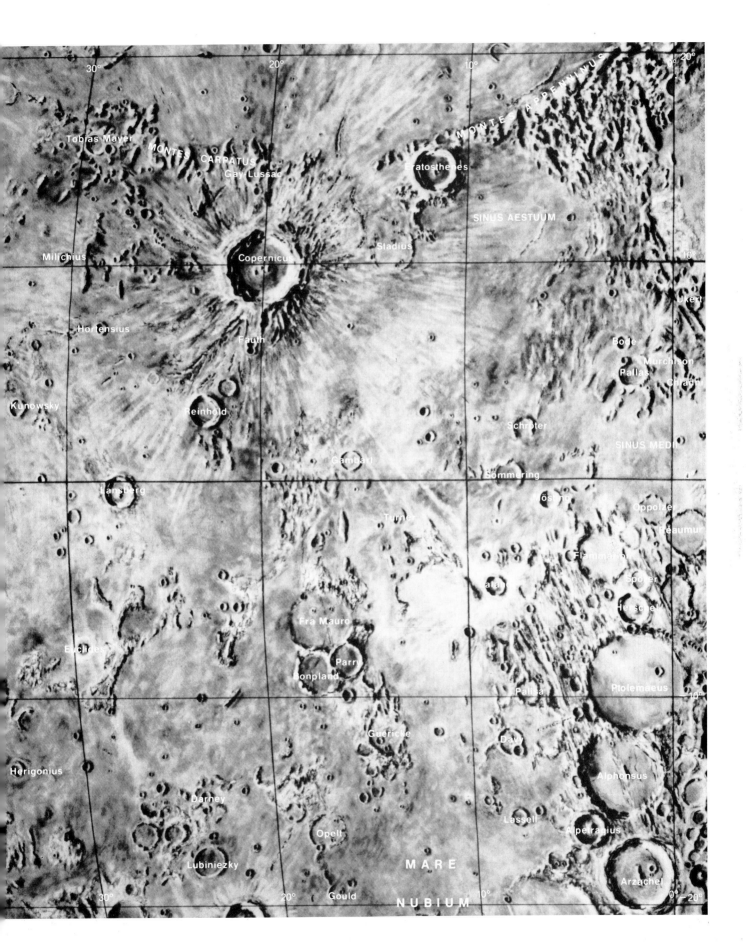

MONTES CARPATUS

MONTES APPENNINUS

Tobias Mayer

Gay Lussac

Eratosthenes

SINUS AESTUUM

Milichius

Stadius

Copernicus

Ukert

Hortensius

Fauth

Bode

Pallas

Murchison

Chladni

Kunowsky

Reinhold

Schröter

SINUS MEDII

Cambart

Sömmering

Lansberg

Mösting

Oppolzer

Réaumur

Turner

Flammarion

Spörer

Herschel

Fra Mauro

Euclides

Parry

Bonpland

Palisa

Ptolemaeus

Guericke

Davy

Herigonius

Alphonsus

Darney

Lassell

Alpetragius

Opelt

Lubiniezky

MARE

Gould

NUBIUM

Arzachel

East Central Region Orthographic

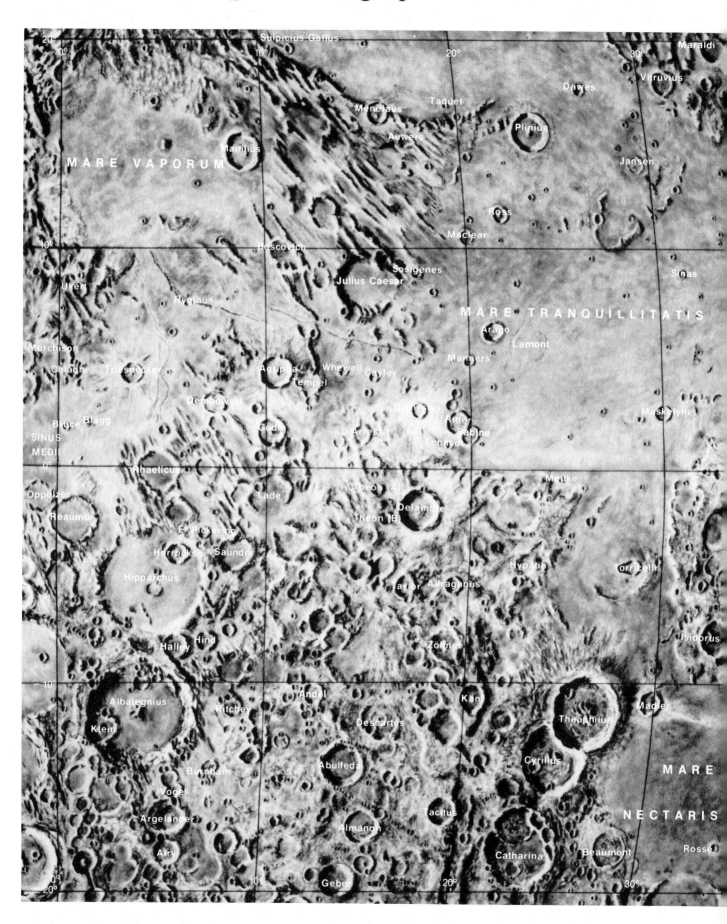

MARE VAPORUM

MARE TRANQUILLITATIS

SINUS
MEDII

MARE

NECTARIS

Sulpicius Gallus
Maraldi
Vitruvius
Dawes
Taquet
Menelaus
Auwers
Plinius
Jansen
Ross
Maclear
Boscovich
Sosigenes
Sinas
Julius Caesar
Ukert
Hyginus
Arago
Lamont
Manners
Murchison
Chladni
Triesnecker
Agrippa
Whewell
Cayley
Tempel
Dembowski
Maskelyne
Bruce
Blagg
Godin
Ritter
D'Arrest
Sabine
Schmidt
Rhaeticus
Moltke
Oppolzer
Lade
Theon (A)
Reaumur
Delambre
Theon (B)
E. Pickering
Horrocks
Saunder
Hypatia
Torricelli
Hipparchus
Taylor
Alfraganus
Halley
Hind
Zöllner
Isidorus
Albategnius
Andel
Kant
Mädler
Ritchey
Theophilus
Klein
Descartes
Cyrillus
Burnham
Abulfeda
Vogel
Tacitus
Argelander
Almanon
Catharina
Beaumont
Rosse
Airy
Gebel

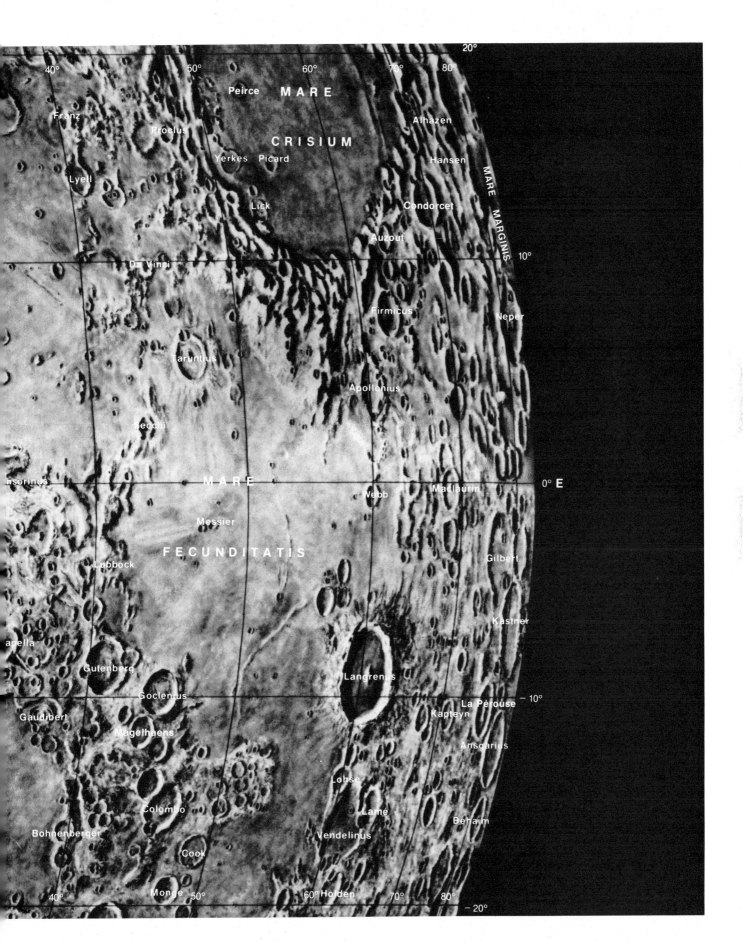

40° 50° 60° 70° 80° 20°

Franz

Peirce **M A R E**

Proclus

C R I S I U M

Lyell

Yerkes Picard

Alhazen

Lick

Hansen

Da Vinci

Condorcet

MARE MARGINIS

Auzout

10°

Firmicus

Neper

Taruntius

Apollonius

Secchi

M A R E 0° E

Webb Maclaurin

Messier

F E C U N D I T A T I S

Gilbert

Lubbock

apella Kästner

Gutenberg

Langrenus

Goclenius − 10°

Gaudibert La Pérouse
Kapteyn

Magelhaens

Ansgarius

Colombo

Lohse

Bohnenberger Lamé Behaim

Cook

Vendelinus

Monge 50° 60° Holden 70° 80° − 20°

63

Central Region

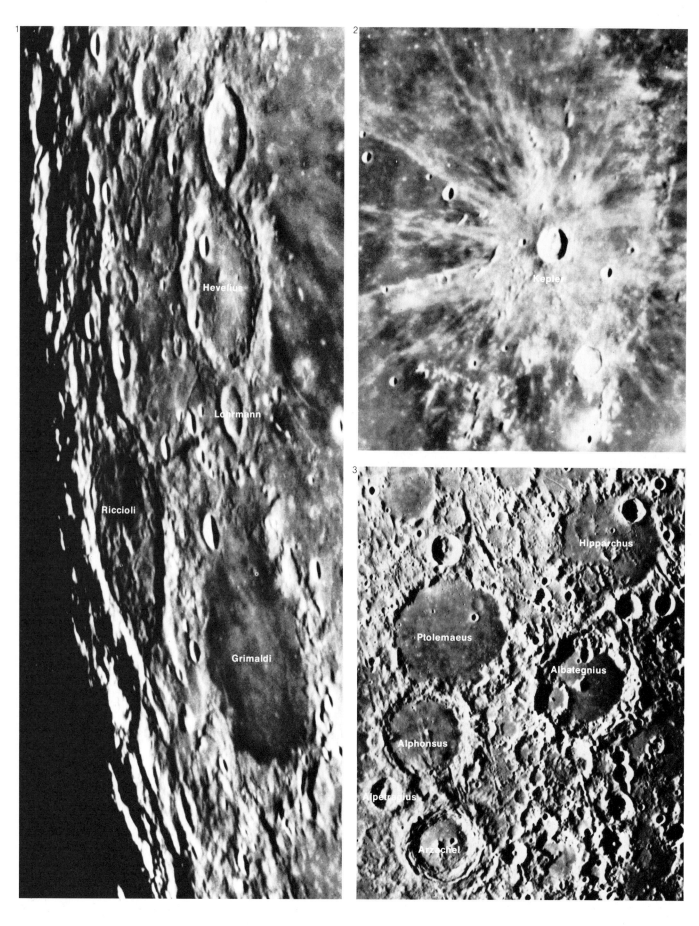

1 Hevelius
Lohrmann
Riccioli
Grimaldi

2 Kepler

3 Hipparchus
Ptolemaeus
Albategnius
Alphonsus
Alpetragius
Arzachel

West central region (left)
1. Key features: Grimaldi (diameter 193 km); Riccioli (160 km); Hevelius (122 km); Lohrmann (45 km). Grimaldi is generally regarded as the darkest region on the Moon; it is a huge walled formation associated with a mascon (*see* page 20). It makes up a pair with another low-walled enclosure, Riccioli, and is included in a long chain with Cavalerius, the convex-floored Hevelius and Lohrmann. Hevelius has a system of rills on its floor.

2. Kepler (diameter 35 km). Kepler is a bright-walled ray-crater with radial interior bands; the walls are so heavily terraced that they seem almost to be double in a number of places.

3. Key features: Ptolemaeus (diameter 148 km); Alphonsus (129 km); Arzachel (97 km); Hipparchus (145 km); Albategnius (129 km); Alpetragius (43 km). The Ptolemaeus chain is one of the most magnificent on the Moon; Alphonsus has a system of rills on its floor and was the site of Kozyrev's reported TLP of 1958. Alpetragius has a massive central mountain with summit craterlet. Hipparchus and Albategnius are also huge, but less well preserved; Hipparchus is difficult to identify under high lighting conditions.

East central region (right)
4. Key features: Mare Crisium (diameter 450 km by 563 km); Picard (34 km); Peirce (19 km); Proclus (29 km); Taruntius (50 km). The Mare Crisium is the most prominent of the seas separated from the main systems; it has a fairly smooth floor, though with various craterlets and two larger formations, Picard and Peirce. Outside it to the west is Proclus, a brilliant crater 2,400 m deep; Taruntius, on the edge of the Mare Fecunditatis, has low, narrow walls, nowhere attaining more than 1,000 m. Taruntius is a good example of a concentric crater.

5. Key features: Theophilus (diameter 101 km); Cyrillus (97 km); Catharina (89 km). Theophilus, with massive, terraced walls and a complex central mountain group, is the most perfect of the three; Cyrillus has a rougher floor and is interrupted by Theophilus, while Catharina includes a low-walled ruined ring and lacks any central peak.

6. Key features: Petavius (diameter 160 km); Vendelinus (160 km); Humboldt (193 km); Holden (40 km); Wrottesley (55 km). Petavius is a magnificent feature, with very complex ramparts; Vendelinus is less prominent and probably older.

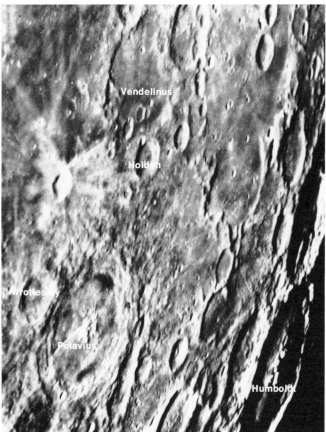

South-west Region Orthographic

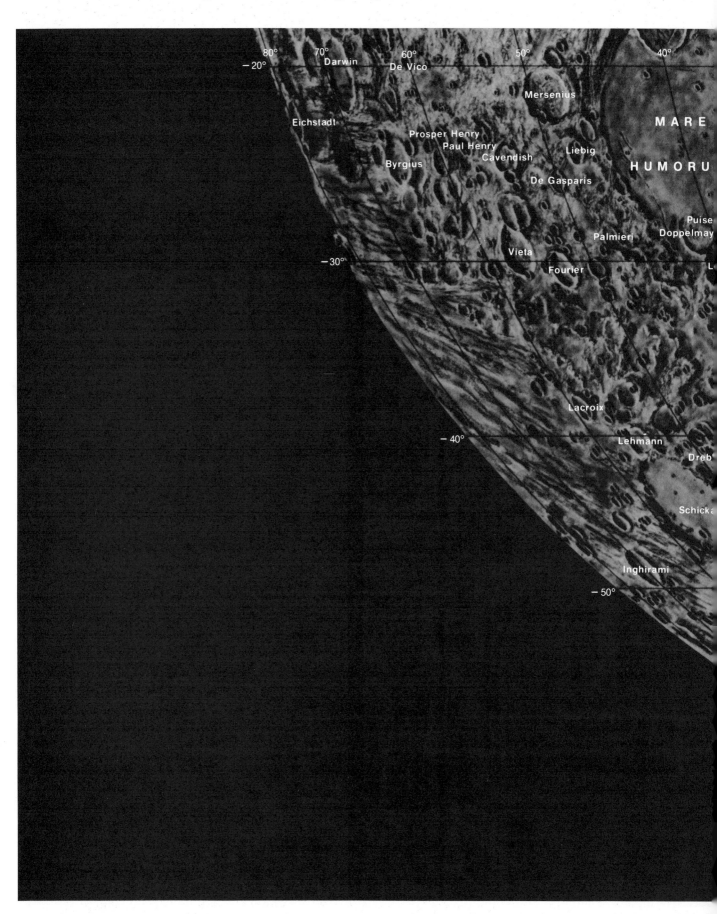

MARE

NUBIUM

30°
Agatharchides
Bullialdus
Gould
20°
10°
20°

Wolf
Nicollet
Birt
Thebit
König
Hippalus
Lippershey
Purbach
Kies
Campanus
Regiomontanus
Mercator
Heiodus
30°
Pitatus
Vitello
Weiss
Deslandres
Hell
Ramsden
Cichus
Walter
Lapaute
Capuanus
Wurzelbauer Gauricus
Elger
Lexell
Ball
ausius
Hauet
Renato
Emley
Miller
Haidinger
Heinsius
Sasserides
40°
Hainzel
Orontius
Nasireddin
Epimenides
Barker
Huggins
Mee
Tycho
Pictet
Saussure
Wilhelm
Street
Proctor
Nöggerath
Longomontanus
50°
Vargentin
Maginus
Nasmyth
Schiller
Bayer
Phocylides
Deluc
Rost
Porter
Clavius
Segner
60°
60°
Zucchius
Scheiner
Rutherfurd
Bettinus
Blancanus
Bailly
Cysatus
Gruemberger
Kircher
Klaproth
70°
Wilson
Moretus
Casatus
Short
Newton
80°
80°

90°
S

67

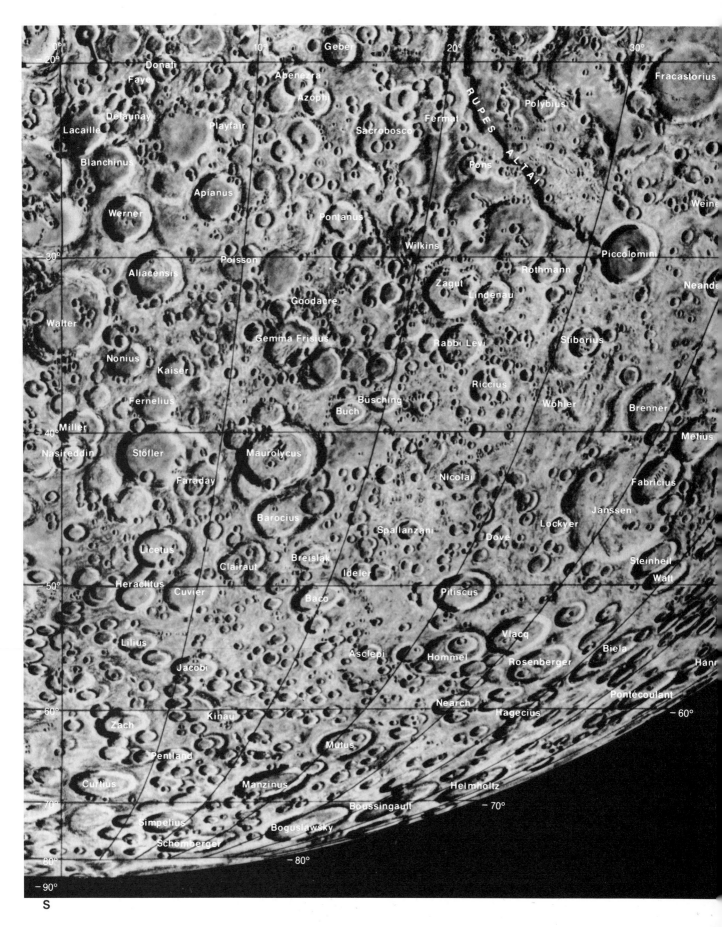

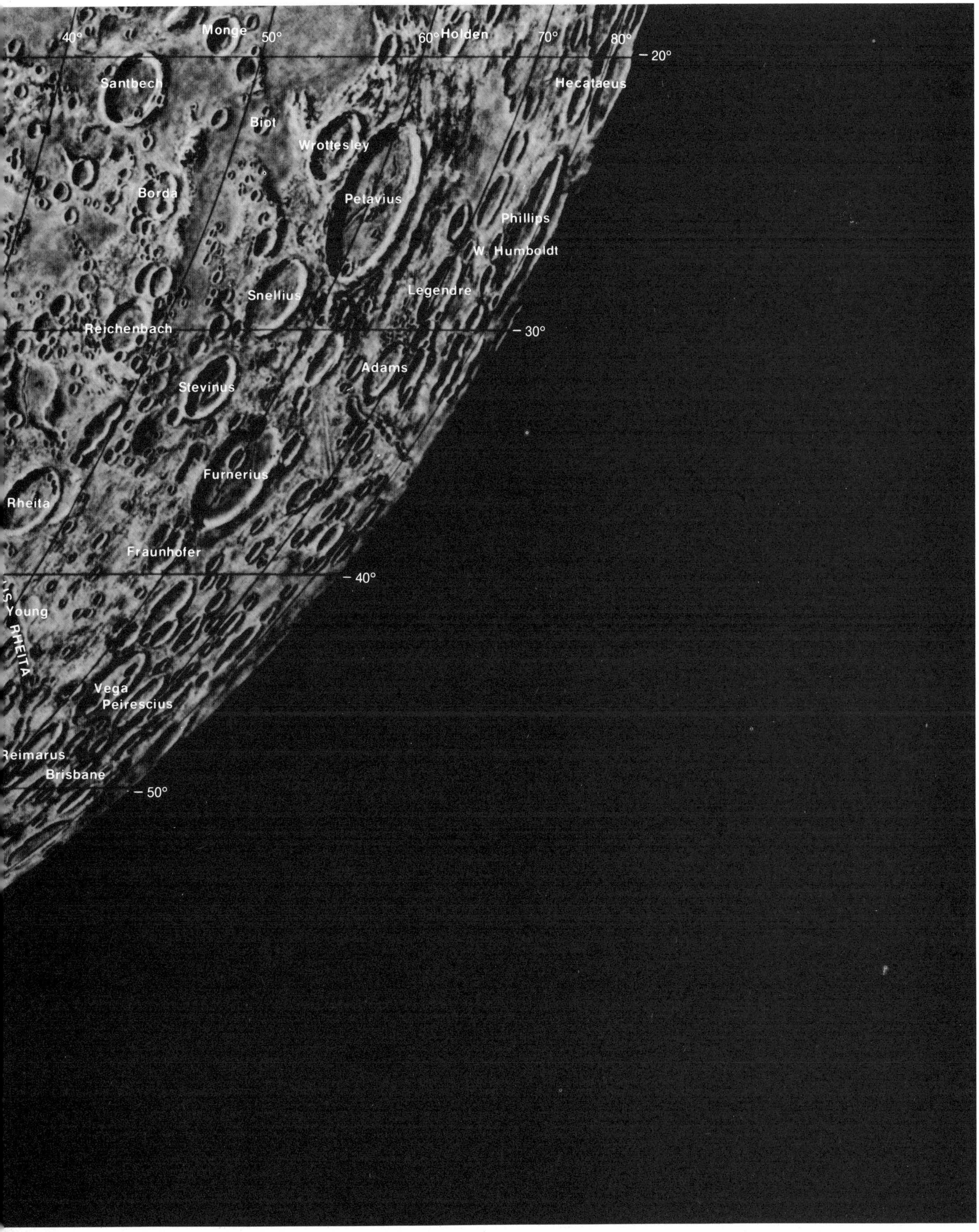

40° Monge 50° 60° Holden 70° 80° — 20°

Santbech   Hecataeus

Biot

Wrottesley

Borda   Petavius

Phillips

W. Humboldt

Snellius   Legendre

Reichenbach   — 30°

Adams

Stevinus

Furnerius

Rheita

Fraunhofer

— 40°

S. Young

RHEITA

Vega
Peirescius

Reimarus

Brisbane

— 50°

South Region

Gassendi

Mersenius

MARE HUMORUM

Hippalus

Campanus

Mercator

Doppelmayer

Vitello

South-west region (left)
1. Key features: Mare Humorum; Gassendi (diameter 89 km); Mersenius (72 km); Doppelmayer (68 km); Vitello (38 km); Hippalus (61 km); Mercator and Campanus (each 38 km). The Mare Humorum is very well-marked, with mountainous borders, except in the north, where there is a break between Gassendi and the rather ill-formed Agatharchides (diameter 48 km). Doppelmayer and Hippalus have had their seaward walls almost destroyed, so that they have become bays, although each has the remnant of a central peak; Hippalus is associated with a major system of rills. Rills are also associated with the convex-floored Mersenius. Vitello is an excellent example of a concentric crater. Gassendi is a superb formation, with its seaward wall reduced but still continuous and there are many rills on its floor. It is one of the lunar spots most prone to TLP.

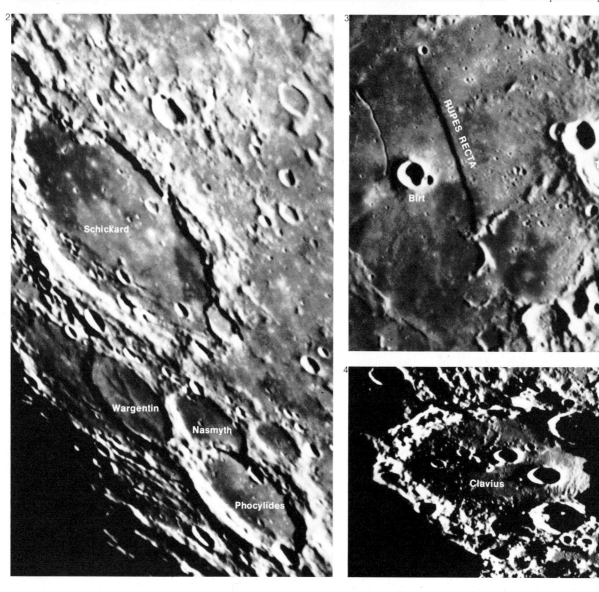

Schickard

Wargentin

Nasmyth

Phocylides

RUPES RECTA

Birt

Theob

Clavius

2. Key features: Schickard (diameter 202 km); Wargentin (89 km); Phocylides (97 km); Nasmyth (74 km). Schickard has a darkish floor and moderate walls; it is one of the largest walled plains on the Moon. Wargentin is the famous plateau, filled with lava almost to the brim.

3. Key features: Rupes Recta; Thebit (diameter 60 km); Birt (18 km). Rupes Recta (the "Straight Wall") is not, in fact, a wall, nor is it straight; it is the best example of a lunar fault. Thebit is interrupted by a smaller crater, which is in turn interrupted by a third.

4. Clavius (diameter 232 km). A massive-walled enclosure, one of the very largest on the Moon; the floor contains a curved chain of craters and rises to over 3,600 m. The walls are interrupted by two craters, Rutherfurd and Porter, each 40 km in diameter.

South-east region (right)

5. Fracastorius (diameter 97 km). Sometimes known simply as Fracastor. It leads off the Mare Nectaris, and its seaward wall has been so reduced that it is barely traceable, turning Fracastorius into a bay. The floor contains a considerable number of small craterlets.

6. Altai Scarp—formerly but much less accurately known as the Altai Mountains. It forms part of the ring system of the Mare Nectaris. It rises to an average of 1,800 m above the general level to the east, but much less to the west.

7. Vallis Rheita. This is one of the most significant features of its type on the Moon. It is close to the well-formed crater Rheita (diameter 68 km) and has a total length of about 185 km; the breadth across its widest point is 24 km. It is not a true valley but is made up of a chain of craters.

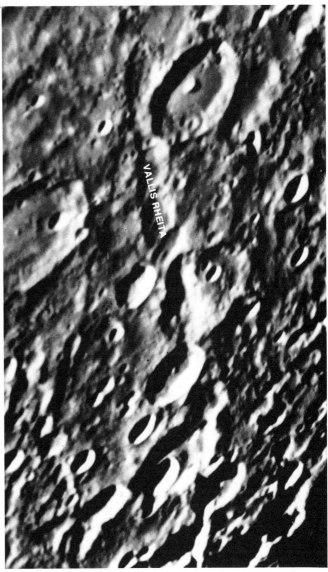

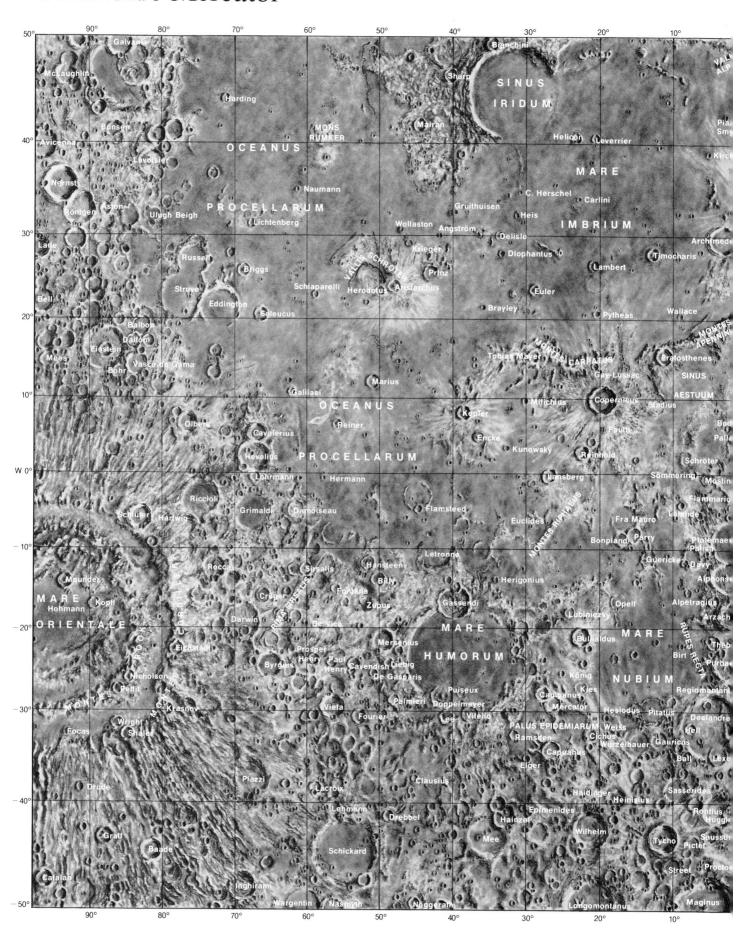

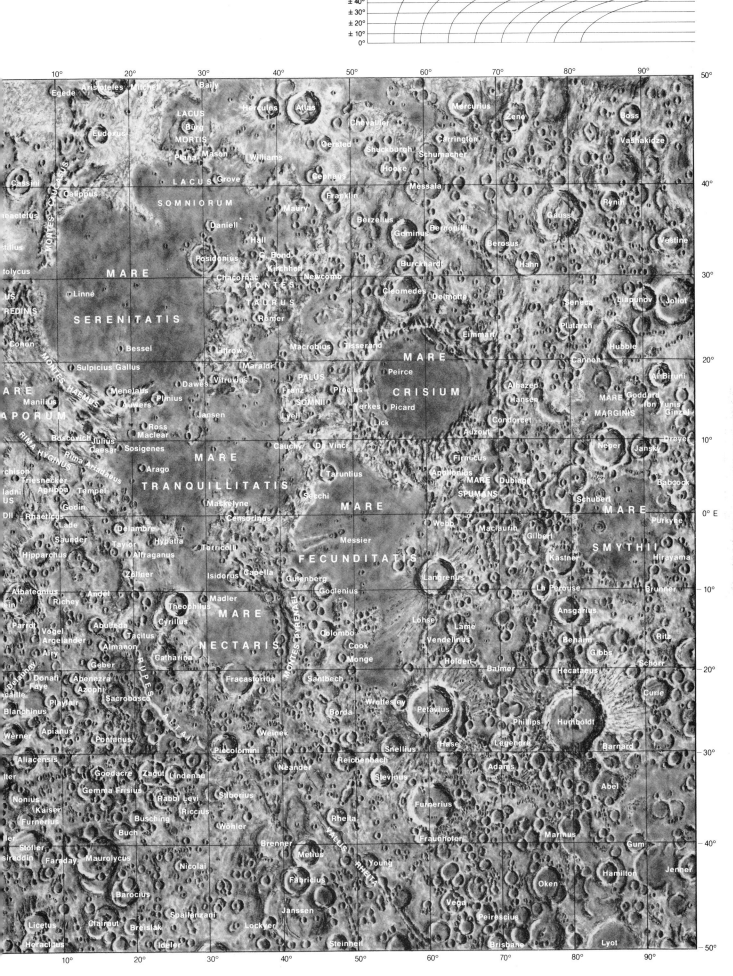

Egede Aristoteles Mitchell Baily
MONTES CAUCASUS LACUS Bürg Hercules Atlas Chevallier Mercurius Zeno Boss
Eudoxus MORTIS Oersted Shuckburgh Carrington Vashakidze
Cassini Plana Mason Williams Hooke Schumacher Rynin
Calippus LACUS Grove Cepheus Messala Gauss Vestine
reaetetus SOMNIORUM Maury Franklin Berosus
tilus Daniell Berzelius Bernouilli Hahn
olycus Hall Geminius
MARE Posidonius G. Bond Burckhardt
REDINIS Linné Chacornac Kirchhoff Cleomedes Delmotte Seneca Liapunov Joliot
US MONTES Newcomb Plutarch Hubble
Cohon SERENITATIS TAURUS Römer Eimmart Cannon
Bessel Littrow MARE Al Biruni
ARE Sulpicius Gallus Maraldi CRISIUM Peirce Alhazen MARE Goddard
APORUM MONTES HAEMUS Menelaus Vitruvius PALUS Hansen MARGINIS Ibn Yunus Ginzel
Manilius Dawes Franz Proclus Yerkes Picard Condorcet Dreyer
RIMA HYGINUS Plinius SOMNII Lyell Lick Auzout Neper Jansky
Boscovich Julius Jansen Cauchy Di Vinci Firmicus Babcock
rchison Triesnecker Caesar Ross Taruntius Apollonius MARE Dubiago
ladni Agrippa Rima Ariadaeus Maclear MARE SPUMANS Schubert
US Tempel Arago TRANQUILLITATIS Secchi MARE MARE
DII Godin Maskelyne Webb Maclaurin SMYTHII Purkyne
Rhaeticus Lade Censorinus FECUNDITATIS Gilbert Hirayama
Hipparchus Delambre Hypatia Messier Kästner
Saunder Taylor Torricelli Brunner
Albategnius Andel Alfraganus Isidorus Capella Gutenberg Langrenus La Pérouse
ein Richey Madler Goclenius Ansgarius
Parrot Vogel Abulfeda Theophilus MARE Colombo Lohse Lame Benaim Ritz
Argelander Tacitus Cyrillus NECTARIS Lame Vendelinus Gibbs Schorr
Airy Almanon Catharina Cook Holden Hecataeus
Geber Monge Balmer Curie
Delaunay Abenezra Fracastorius Santbech Petavius Humboldt Barnard
aille Donati Azophi MONTES PYRENAEI Wrottesley Phillips
Faye Sacrobosco Borda Snellius Hase Legendre
Playfair Weinek Balmer
Blanchinus Apianus Piccolomini Reichenbach Adams Abel
Werner Pontanus Neander Stevinus
Aliacensis RUPES ALTAI Furnerius
lter Goodacre Zagut Lindenau Silberius Rheita Fraunhofer Marinus
Nonius Gemma Frisius Rabbi Levi Riccius Wöhler Gum
Kaiser Büsching Brenner VALLIS RHEITA Young Hamilton Jenner
Furnerius Buch Metius Oken
ller Stöfler Maurolycus Nicolai Fabricius Vega
sireddin Faraday Barocius Janssen Peirescius
Licetus Clairaut Breislak Spallanzani Lockyer Brisbane Lyot
Heraclitus Ideler Steinheil

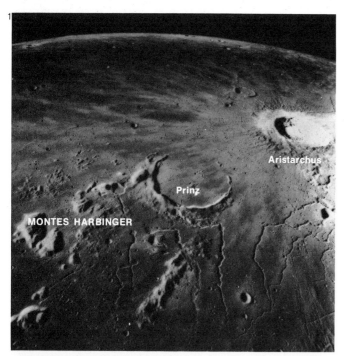

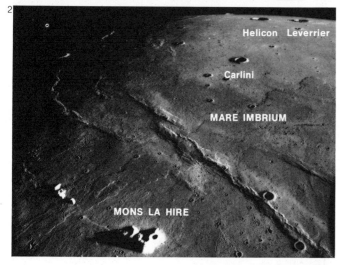

North-west region (left)

1. Key features: Aristarchus (diameter 37 km), Prinz (50 km); Montes Harbinger. This is one of the most fascinating regions of the Moon. Prinz is an incomplete crater lacking its southern (upper) wall. The Harbinger Mountains are really clumps of hills rather than a true range, and the whole region is rich in rills. (Apollo 15)

2. Key features: Mare Imbrium area; Peak La Hire; Carlini (diameter 8 km), Helicon (29 km); Leverrier (25 km). The mountain mass La Hire is 1,500 m high with a summit craterlet. Beyond the ridge is the well-marked bright crater Carlini, the twin craters Helicon (left) and Leverrier (right), also shown on the Earth-based photograph on page 58. (Apollo 15)

3. Lambert (diameter 29 km). Situated in the Mare Imbrium, this is a well-formed crater with walls which are massive and terraced, though not particularly bright. When this picture was taken, much of the interior of Lambert was in shadow, concealing the central formation—which is a crater instead of a mountain mass. Adjoining Lambert, below it on the picture, is a large "ghost ring". (Apollo 15)

4. Aristarchus (diameter 37 km). The central peak, to the upper left in the picture, casts a pronounced shadow; the floor of the crater is rough, and the wall is of great complexity. (Lunar Orbiter 5)

5. Copernicus (diameter 97 km). This great ray-crater has massive, terraced walls and a regular outline. The wall is well-defined, with a sharp crest. The floor of the crater is pitted with craterlets, and is evidently sunken; the wall rises only to a modest height above the surface outside the formation. (Lunar Orbiter 5)

South-west region (right)

6. Key features: Mare Nubium; Fra Mauro (diameter 81 km), Bonpland 58 km), Parry (42 km). Note the "rolling rock" in Fra Mauro. This was the intended landing-area of the ill-fated Apollo 13; Apollo 14 did in fact land in the region, taking astronauts Shepard and Mitchell onto the surface. (Apollo 16)

7. Key features: part of Ptolemaeus (diameter 148 km), Alphonsus (129 km), Arzachel (97 km). The floor of Ptolemaeus is rich in craterlets, but there is no central peak. (Apollo 16)

8A and B. Tycho (diameter 84 km). This great ray-crater is one of the most perfect of the walled formations; the rays are seen only under high illumination, and so are not shown here. The walls are high and terraced, and there is a central mountain complex. The area enclosed in the rectangle is shown in more detail on the right-hand picture, and the roughness of the floor leading up to the base of the inner wall is very evident. (Lunar Orbiter 5)

9. Gassendi (diameter 89 km). Like many other areas rich in rills, Gassendi has been the site of a number of reported TLP (see page 20). (Lunar Orbiter 5)

10. Key features: Rima Sirsalis; Sirsalis (diameter 32 km), Crüger A (22 km), De Vico A (30 km), De Vico T (35 km), De Vico (24 km). Sirsalis intrudes into its "twin", Sirsalis A. The rill runs past Crüger A and cuts the irregular De Vico A. (Lunar Orbiter 4)

11. Key features: Hippalus area; Hippalus (diameter 61 km), Agatharchides A (48 km), Campanus (38 km), Mercator (38 km). Hippalus has a central peak and is associated with a fine system of curved, parallel rills, (Lunar Orbiter 4)

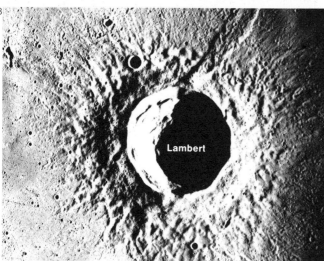

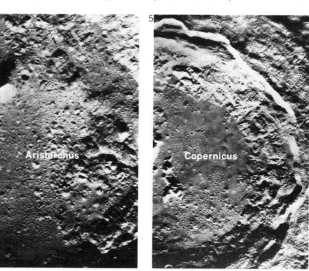

Near Side East Region

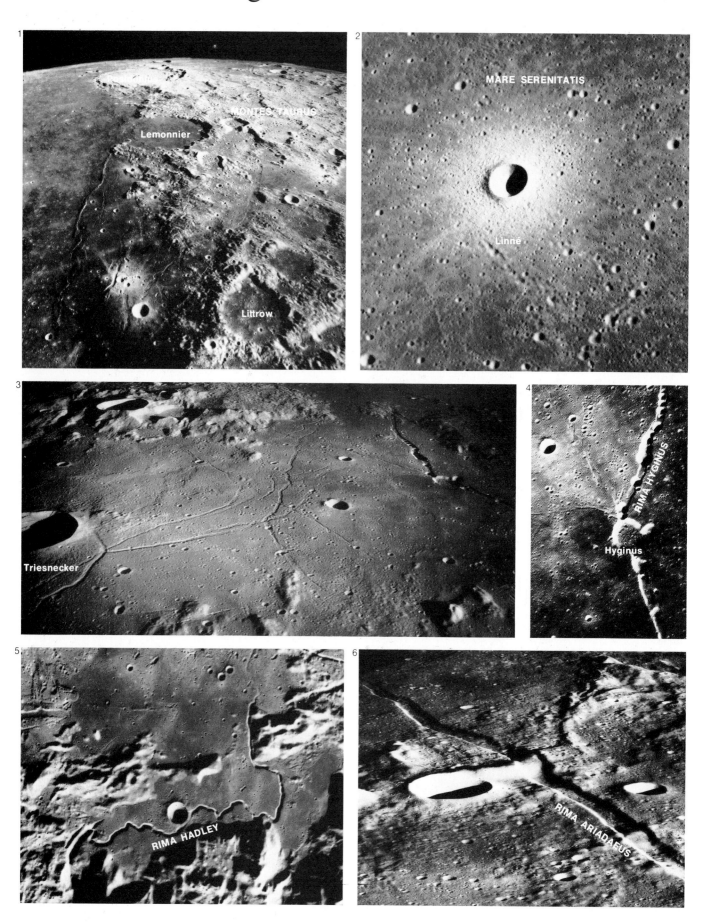

1

MONTES TAURUS

Lemonnier

Littrow

2

MARE SERENITATIS

Linné

3

Triesnecker

4

RIMA HYGINUS

Hyginus

5

RIMA HADLEY

6

RIMA ARIADAEUS

North-east region (left)

1. Key features: Posidonius (diameter 96 km), Lemonnier (55 km), Littrow (35 km); Taurus Mountains. The photograph shows part of the Mare Serenitatis. It was in this general area that the Soviet Lunokhod 2 crawled around in 1973, and where it remains. The so-called Taurus Mountains do not make up a connected range. Apollo 15 landed in the Taurus–Littrow area in December 1972. (Apollo 17)

2. Linné (diameter 4 km). This is a good example of a fresh, sharp-edged crater of probable impact origin. It is located on the western region of the Mare Serenitatis. (Apollo 15)

3. Triesnecker (diameter 23 km). Triesnecker is regular and notable mainly because of its association with a system of rills, which are not crater-chains—unlike the so-called Hyginus Rill, which appears to the upper right. This region lies between the Mare Vaporum and the Sinus Medii. (Apollo 10)

4. Key features: Rima Hyginus; Hyginus (diameter 6 km). The Hyginus Rill is essentially a crater-chain. The regular crater above Hyginus on the picture is Hyginus B. (Lunar Orbiter 5)

5. Rima Hadley. This photograph shows a region near the edge of the Mare Imbrium; it was near here that astronauts Scott and Irwin landed, and drove in their Lunar Roving Vehicle almost to the edge of the rill. (Apollo 15)

6. Rima Ariadaeus. Unlike the Hyginus formation, Ariadaeus is a genuine rill and not a crater-chain. It is over 240 km long. (Apollo 10)

South-east region (right)

7. Rupes Altai. This feature, also known as the "Altai Scarp", forms part of the ring system associated with the Mare Nectaris. It rises to a greater height above the surrounding region on the east than it does above the region to the west. The scarp has a relatively low crater density. (Lunar Orbiter 4).

8. The Messier twins (diameters approximately 13 km). This is a strange pair of craterlets in the Mare Fecunditatis. Messier A was formerly known as W. H. Pickering. From it extends the unique "comet" ray towards the western edge of the Mare. The craters show marked apparent changes at different conditions of illumination, but suggestions of real structural changes in historic times may safely be dismissed. (Apollo 15)

9. Key features: Stöfler (diameter 145 km), Faraday (64 km). Part of Stöfler has been destroyed by the intrusion of Faraday. (Lunar Orbiter 4)

10. Langrenus (diameter 137 km). Langrenus is a vast walled plain with high terraced walls rising to 2,700 m above the sunken floor. (Apollo 15)

11. Theophilus (diameter 101 km). The terraced inner ramparts rise to over 5,000 m above a floor which contains a massive central group of elevations. (Apollo 16)

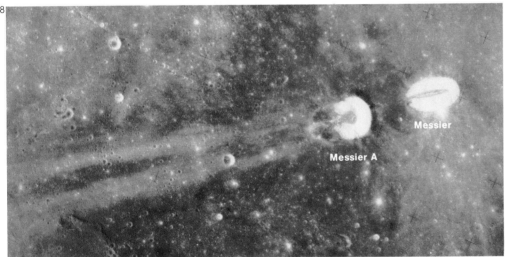

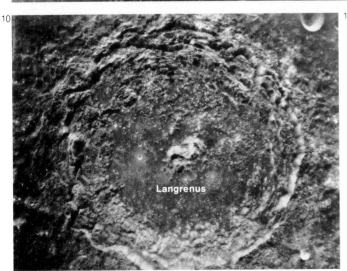

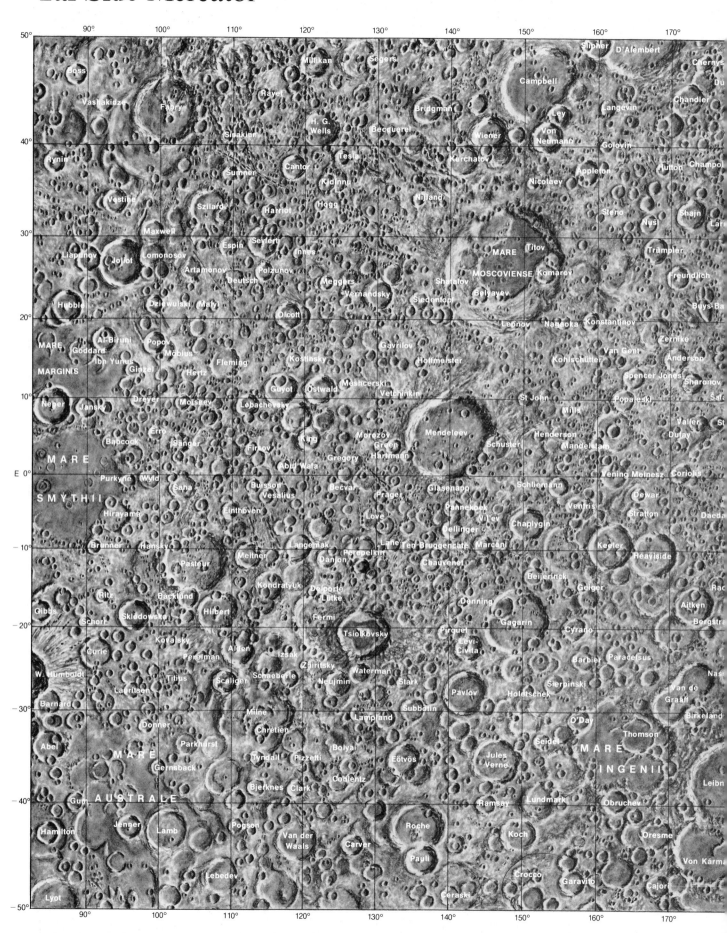

	0	100	200	300	400	500	600	700	800 km
± 50°									
± 40°									
± 30°									
± 20°									
± 10°									
0°									

170° 160° 150° 140° 130° 120° 110° 100° 90°   50°

Paraskevopoulos
Montgolfier   Esnault
Woltjer   Stoletov   Pelterie   Schlesinger   Sarton   Chapman   Galvani
        Schneller   Kulik   Gullstrand   Stefan   McLaughlin
Ehrlich   Fowler   Wood   Wegener   Schönfeld
nkler   Von Zeipel   Quételet   Bragg
                                                     Landau   Lacchini   Bunsen
                                        Thiel   Avicenna
Moore   Parsons   Rasumov
                        Gadomski   Charlier   Frost   Petropavlovsky
        Evershed   Klute   Nernst
                    Sanford   Winlock   Lorentz   Röntgen
Cockcroft Van den Bergh Blazhko Teisserenc   Herley   Aston
Fitzgerald   Kovalevskaya   Laue
                    Joule   Bobone   Leucippus
Morse   Mineur   Parenago   Berkner   Helberg   Balboa
        Jackson   McNally   Comrie   Bell   Dalton
Hayford   Harvey   Comstock Sternberg Robertson Einstein
        McMath Bredikhin Mitra Mach   Poynting Fersman Weyl Ohm Alter Nobel Vasco
                                Kekulé   Kammerlingh Onnes Mees Bohr da Gama
        Raimond   Pease
                    Henyey   Kolhörster
            Dirichlet   Elvey
                        Artem'ev   Kuo Shou Ching
Zhukovsky Lebedinsky   Tsander   Michelson
        Englehardt   Leuschner
Krasovsky   Kibalchich Chaucer   Schlüter
Congreve   Yavilov   Hertzsprung   Grachev
                                                                                        0° W
Icarus   Korolev   Ingalls Timiryazev   Lucretius
                    Krylov   Sechenov   10°
Amici Crookes   Evans Kearons   Lowell
        Doppler   Das   Metchnikoff Van Gu   Friedman   Maunder
McKeller   Galois Paschen   Joffe   MARE Kopff
        Mohorovicic   Lodygin   Belopolsky Houzeau   Hohmann
De Vries   Lewis   ORIENTALE
        Sniadecki   Wilsing   Strömgren   Golitzyn
Orlov   Plummer   Von der Pahlen Gerasimovic   Nicholson
                    Walker   Ellerman   Pettit
euwenhoek Rumford   Barringer   Mariotte
                                                                              Wright
Davisson Oppenheimer Dryden   Kleimenov Chebyshev   Steklov   Shaler
                    Apollo   Boakan Langmuir Brouwer   Drude
            Chaffee   Lovell   Stetson   Chant
Maksutov   Anders   Buffon   Fenyi   Graff
Nishina   Grissom   Leavitt   Rydberg Cetalan
        Hendrix   White   Mendel   Guthnick
                        Riedel

170° 160° 150° 140° 130° 120° 110° 100° 90°   50°

Far Side East Region

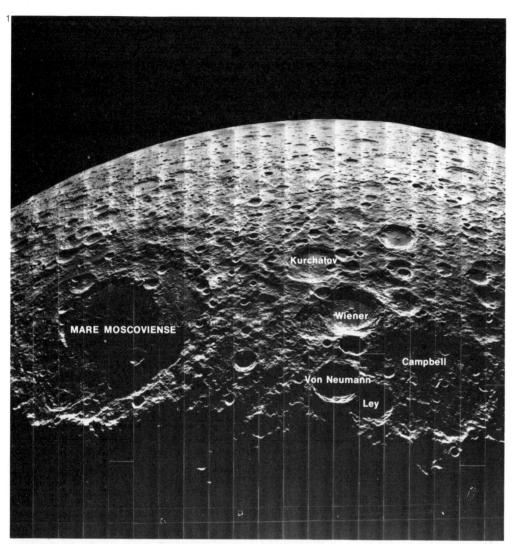

North-east Region

North-east Region

1. Key features: Mare
Moscoviense; Campbell (diameter
260 km), Ley (80 km), Von
Neumann (110 km), Wiener
(140 km), Kurchatov (110 km).
The Mare Moscoviense was one
of the first features to be
identified on the Moon's far side;
it was shown as a dark patch on
the Luna 3 pictures of 1959. It is
certainly not a major mare, but it
does have a dark floor, and there
is an outer ring. This whole area
is very rough. (Lunar Orbiter 5)

2. Key features: Compton
(diameter 175 km), Fabry
(210 km), Szilard (140 km), Seyfert
(110 km), Cantor (100 km), H. G.
Wells (160 km), Millikan (140 km).
This is another very rough highland area, lying on the far side of
the Moon towards the north pole,
beyond the Mare
Humboldtianum. Compton is of
special interest in as much as it
has both a central elevation and
an inner ring of peaks—a very
rare combination. (Lunar Orbiter
5)

3. Key features: Lobachevsky
(diameter 80 km), Guyot (110 km)
Ostwald (120 km), Fleming
(130 km), King (90 km), Olcott
(70 km). This region lies on the
far hemisphere beyond Mare
Smythii. It contains some well-
formed craters; for instance
Lobachevsky has fairly regular
walls (slightly polygonal in
outline) and a group of central
elevations. King also is well-
formed, with terraced walls; a
low-walled crater makes up a
chain with King and Ostwald.
Terracing is also noticeable in the
crater Olcott. (Apollo 16)

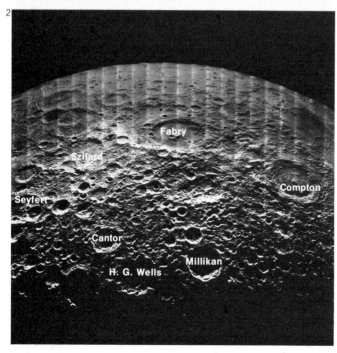

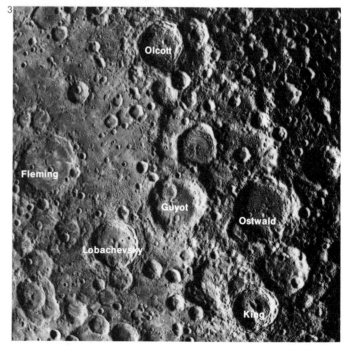

South-east region

4. Key features: Gagarin (diameter 240 km), Pavlov (130 km), Jules Verne (130 km), Mare Ingenii (350 km), O'Day (80 km), Thomson (100 km), Van de Graaff (210 km), Aitken (130 km), Heaviside (140 km), Keeler (140 km). This rough region includes the so-called Mare Ingenii, an irregular formation with a darkish floor; it has no real claim to the title of "mare". Much the more interesting feature is the crater Van de Graaff. It is in many ways unique; it is a depression about 4 km deep, and is associated with a strong magnetic anomaly, possibly indicating buried volcanic rock. Van de Graaff also shows greater radioactivity than its surroundings. (Lunar Orbiter 2)

5. Key features: Tsiolkovsky (diameter 198 km), Fermi (210 km), Alden (90 km), Milne (250 km), Waterman (70 km), Neujmin (110 km), Roche (190 km), Pauli (100 km), Pavlov (130 km), Delporte (50 km), Lutke (50 km), Chrétien (90 km), Scalinger (100 km). This region adjoins the area shown in photograph 4 (Pavlov appears on both). (Lunar Orbiter 3)

6. O'Day; this is a fairly prominent crater adjoining Mare Ingenii. (Lunar Orbiter 2)

7. Key feature: Tsiolkovsky (diameter 190 km). This formation, one of the most remarkable on the Moon, has about half the diameter of the Mare Crisium. Part of the floor is very dark, and obviously consists of mare material. (Apollo 15)

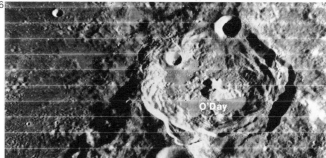

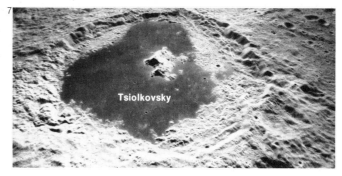

Far Side West Region

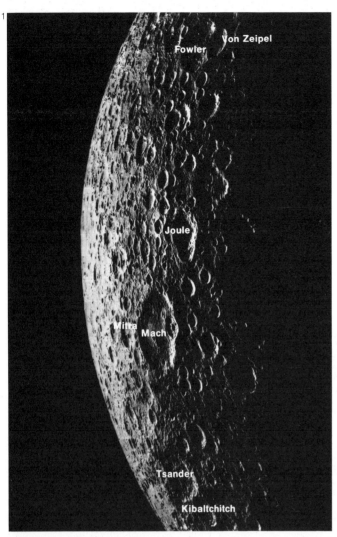

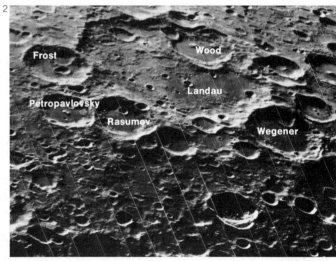

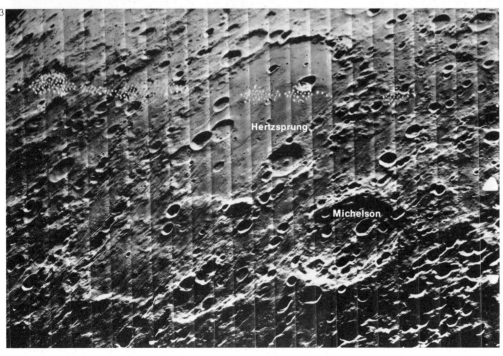

1. Key features: Fowler (diameter 140 km), Von Zeipel (70 km), Joule (120 km), Mitra (80 km), Mach (180 km), Tsander (140 km), Kibaltchitch (100 km). Fowler has a darkish floor containing some small craterlets; its western wall is broken by a small formation, Von Zeipel. Tsander and Kibaltchitch make up a pair; they lie between two huge enclosures, Hertzsprung to the west on the unilluminated part of the Moon and Korolev to the east (not shown in this photograph). (Lunar Orbiter 5)

2. Key features: Landau (diameter 240 km), Wood (110 km), Wegener (100 km), Frost (80 km), Petropavlovsky (80 km), Rasumov (80 km). Landau is considerably deformed by intruding craters of which the most prominent is Wood. Wegener has a somewhat asymmetrical elevation on its floor; it touches Landau to one side, and on the other it abuts on Stefan (not shown in this photograph). (Lunar Orbiter 5)

3. Key features: Hertzsprung (diameter 440 km), Michelson (130 km). Hertzsprung is one of the major features of the Moon's far side. It has an inner section which is darker than the outer part of the interior; the outer ring is broken by Michelson to the north-west and Vavilov the south-east. (Lunar Orbiter 5)

4. Woltjer (diameter 130 km). This and its companion Montgolfier (not shown in this photograph) lie to the east of Fowler, which is shown in photograph 1 on this page. (Lunar Orbiter 5)

South-west region

5. Key features: Mare Orientale (diameter 900 km); Schlüter (80 km), Einstein (150 km). The Mare Orientale is one of the most important features of the Moon. Its basin is comparable in size with that of the Mare Imbrium, but contains little dark lava except near its center. Its outer ring is made up of the Rook Mountains, which are observable from Earth. This photograph also shows Grimaldi and Riccioli, which belong to the near side of the Moon. Schlüter is well-marked with a central peak. Einstein (named "Caramuel" on some of the older maps) has a central crater, high walls and a long ridge on its floor; from the Earth it is visible on the limb only under ideal conditions. (Lunar Orbiter 4)

6. Key features: Korolev (diameter 360 km), Krylov (50 km), Doppler (90 km). Korolev is the second of the trio of huge ringed formations, the others being Hertzsprung and Apollo. Korolev has two named interior craters, Krylov and Ingalls. (Lunar Orbiter 1)

7. Krylov (diameter 50 km). A regular, well-marked crater with a central peak, inside Korolev. A crater-row runs west from it to the wall of Korolev. (Lunar Orbiter 1)

8. Key features: Apollo (diameter 520 km), Barringer (60 km), Dryden (50 km), Chaffee (60 km), Lovell (30 km). Chaffee is one of a curved line of formations lying on the inner ring of Apollo. (Lunar Orbiter 5)

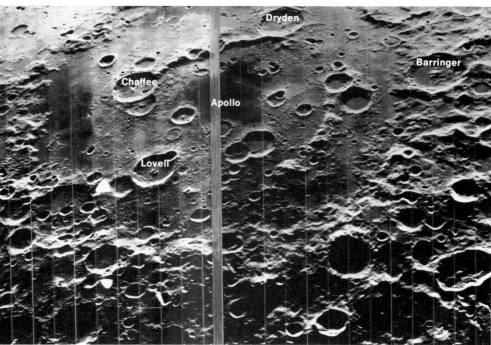

North and South Poles Stereographic

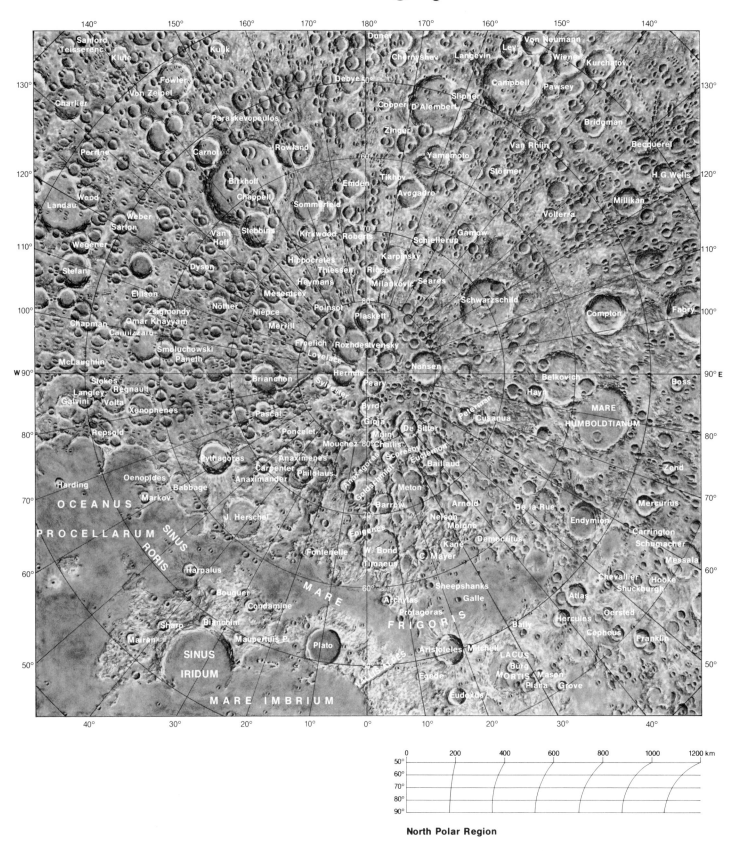

Samford
Teisserenc
Klute
Kulik
Fowler
Von Zeipel
Charlier
Paraskevopoulos
Rowland
Perrine
Carnot
Birkhoff
Chappell
Wood
Weber
Landau
Sarton
Van
Wegener
Hoff
Stebbins
Stefan
Dyson
Ellison
Zsigmondy
Nöther
Chapman
Omar Khayyam
Cannizzaro
Smoluchowski
McLaughlin
Paneth

Duner
Chernyshev
Langevin
Ley
Von Neumann
Wiener
Kurchatov
Debye
Slipher
Campbell
Pawsey
Copper
D'Alembert
Bridgman
Zinger
Becquerel
Van Rhijn
Emden
Tikhov
Störmer
H.G. Wells
Avogadro
Millikan
Sommerfeld
Volterra
Kirkwood
Roberts
Schjellerup
Gamow
Karpinsky
Schwarzschild
Compton
Fabry
Hippocrates
Thiessen
Ricco
Heymans
Milankovic
Seares
Mesemtsev
Niepce
Poinsot
Plaskett
Merrill
Froelich
Rozhdestvensky
Lovelace
Hermite
Nansen

Stokes
Brianchon
Peary
Belkovich
Boss
Regnault
Langley
Sylvester
Hayn
Galvini
Volta
Byrd
Peterman
MARE
Xenophenes
Pascal
Gioja
Cusanua
HUMBOLDTIANUM
Repsold
Poncelet
Main
De Sitter
Zeno
Mouchez
Challis
Pythagoras
Anaximenes
Scoresby
Baillaud
Carpenter
Philolaus
Euctemon
Mercurius
Harding
Anaximander
Goldschmidt
Meton
Oenopides
Babbage
Endymion
Markov
Barrow
Arnold
Carrington
OCEANUS
J. Herschel
Nelson
De la Rue
Schumacher
PROCELLARUM
Epigenes
Moigno
Messala
SINUS
Fontenelle
Kane
Democritus
Chevallier
Hooke
RORIS
W. Bond
C. Mayer
Atlas
Shuckburgh
Harpalus
Timaeus
Oersted
Bouguer
MARE
Sheepshanks
Hercules
Condamine
Archytas
Galle
Bally
Cepheus
Sharp
Bianchini
Protagoras
Franklin
Mairan
Maupertuis P.
Plato
FRIGORIS
Aristoteles
Mitchell
LACUS
SINUS
Egede
Bürg
MORTIS
Mason
IRIDUM
Eudoxus
Plana
Grove

MARE IMBRIUM

0   200   400   600   800   1000   1200 km

North Polar Region

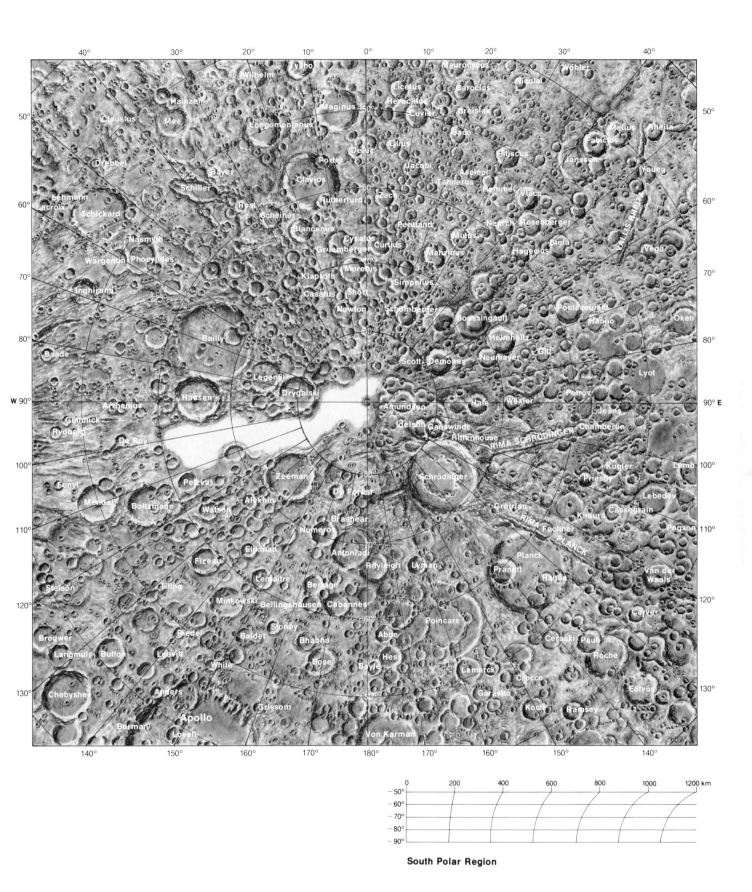

Tycho
Wilhelm
Hainzel
Clausius
Mee
Longomontanus
Maginus
Drebbel
Bayer
Porter
Schiller
Clavius
Rost
Rutherfurd
Scheiner
Blancanus
Lehmann
Schickard
Cysatus
Lacroix
Gruemberger
Curtius
Nasmyth
Klaproth
Moretus
Simpelius
Wargentin
Phocylides
Casatus
Short
Inghirami
Newton
Schömberger
Baade
Bailly
Legentil
Drygalski
Hausen
Arrhenius
Gutmick
Rydberg
De Roy
Fenyi
Petzval
Zeeman
Mendel
Alekhin
De Forest
Boltzmann
Watson
Brashear
Eijkman
Numerov
Fizeau
Antoniadi
Lemaitre
Rayleigh
Lyman
Bergne
Stetson
Minkowski
Cabannes
Bellingshausen
Poincaré
Sloney
Abbe
Brouwer
Riedel
Baldet
Bhabha
Hess
Langmuir
Buffon
Leavitt
Boyle
White
Chebyshev
Anders
Bose
Grissom
Apollo
Borman
Lovell
Von Karman

Maurocycus
Wöhler
Licetus
Nicolai
Barocius
Heraclitus
Breislak
Cuvier
Bado
Metius
Rheita
Fabricius
Lilius
Pitiscus
Janssen
Deluc
Jacobi
Asclepi
Young
Tannerus
Hommel
Vlacq
Zach
Pentland
Nearch
Rosenberger
Mutus
Vega
Manzinus
Biela
Hagecius
Boussingault
Pontécoulant
Hanno
Helmholtz
Oken
Scott
Demonax
Gill
Neumayer
Lyot
Hale
Wexler
Petrov
Amundsen
Jeans
Idelson
Ganswindt
Chamberlin
Rittenhouse
RIMA SCHRÖDINGER
Schrödinger
Kugler
Priestly
Lamb
Lebedev
Gretrian
Cassegrain
RIMA
Fechner
Kimura
Pogson
PLANCK
Planck
Prandtl
Hagen
Van der Waals
Carver
Lamarck
Ceraski
Pauli
Crocco
Roche
Garavito
Eötvös
Koch
Ramsey

VALLIS RHEITA

| | 0 | 200 | 400 | 600 | 800 | 1000 | 1200 km |

South Polar Region

85

Polar Regions

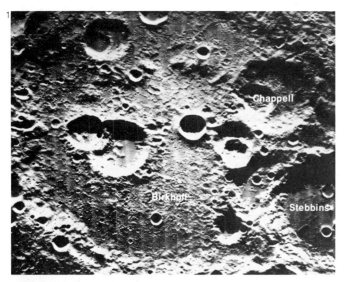

North polar region
1. Key features: Birkhoff (diameter 370 km), Chappell (80 km), Stebbins (130 km). Birkhoff is a vast walled formation with light, rough floors; it is not visible from the Earth. In its interior is an irregular formation, Chappell, which contains an inner crater. The wall of Birkhoff is interrupted by a smaller, higher-walled formation, Stebbins (partially shown here). (Lunar Orbiter 5)

2. Key features: Sinus Roris; Pythagoras (diameter 113 km), Carpenter (65 km), Anaximander (89 km), Babbage (145 km), Oenopides (68 km), Markov (45 km). Pythagoras has high walls and a massive central mountain group; from Earth it is always too foreshortened to be well seen. (Lunar Orbiter 4)

3. Key features: Goldschmidt (diameter 109 km), Anaxagoras (52 km), Epigenes (52 km), Meton (170 km), Barrow (87 km). Goldschmidt is a large crater, interrupted by Anaxagoras; Anaxagoras a deep, well-formed crater; it is very bright, and is a major ray-crater, although the rays are not visible on this picture. Meton is a huge, compound formation with a floor pitted by craterlets. (Lunar Orbiter 4)

4. Vallis Alpes (128 km long). The valley, cutting through the Alps, has a delicate rill running inside it. Unlike, for example, the Rheita Valley (*see* page 71), this feature is a true valley. Above it on the picture is part of the Mare Frigoris, one of the irregular maria running along north of the Mare Imbrium and Mare Serenitatis. (Lunar Orbiter 4)

South polar region

5. Key features: Schrödinger (diameter 310 km), Planck (340 km), Zeeman (200 km). Schrödinger with its associated valley is unfortunately invisible from Earth. (Lunar Orbiter 4)

6. Key features: Clavius (diameter 232 km), Rutherfurd (40 km), Blancanus (92 km), Scheiner (113 km), Schiller (180 km by 97 km), Bayer (52 km), Rost (55 km). Clavius is one of the most impressive and largest walled formations. Schiller is sometimes regarded as a compound formation made up of two or more rings, although it has also been classed as a volcano–tectonic depression. (Lunar Orbiter 4)

7. Key features: Casatus (diameter 104 km), Newton (113 km),

Klaproth (about 100 km). Casatus is the deeper of the pair, and has intruded into Klaproth. They may be seen from Earth. Newton is an exceptionally deep formation. (Lunar Orbiter 4)

8. Key features: Schrödinger (diameter 310 km), Ganswindt (70 km), Idelson (50 km). (Lunar Orbiter 5)

9. Key features: Bailly (diameter 294 km), Legentil (140 km). Bailly has a light floor, irregular walls and has been described as "a field of ruins". In this picture the floor of Legentil is mainly in shadow. (Lunar Orbiter 4)

10. Key features: Pontécoulant (diameter 97 km), Boussingault (78 km), Helmholtz (97 km). The whole of this region is visible from Earth. (Lunar Orbiter 4)

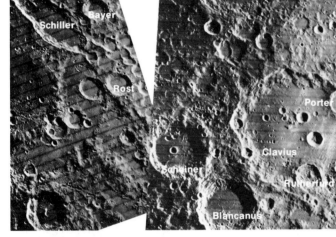

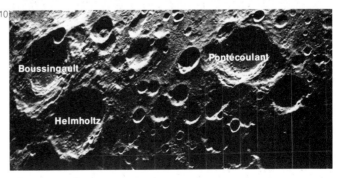

Crater Index

The names of lunar features appearing in the maps are listed with their latitude and longitude; page numbers of the relevant maps are given in the last column. Features other than craters are indicated by SMALL CAPITALS.

Crater Index

Glossary

Albedo The ratio of the amount of light reflected by a body to the amount of light incident on it; a measure of the reflecting power of a body. A perfect reflector would have an albedo of 1. The albedos of the planets are as follows: Mercury 0.06, Venus 0.76, Earth 0.39, Mars 0.16, Jupiter 0.43, Saturn 0.61, Uranus 0.35, Neptune 0.35, Pluto 0.5.

Allotropy The property in a chemical element of existing in different forms, with distinct physical properties but capable of forming identical chemical compounds. Ozone, for example, is an allotropic form of oxygen.

Altitude In astronomy, the angular distance of a celestial body from the horizon. In conjunction with a measurement of AZIMUTH, it describes the position of an object in the sky at a given moment.

Aphelion The point or moment of greatest distance from the Sun of an orbiting body such as a planet. The opposite of PERIHELION.

Asteroid One of a large number of rocky bodies, smaller than a planet but larger than a METEORITE, in orbit around the Sun. Also known as "minor planets", over 99 percent of the asteroids in the SOLAR SYSTEM lie in a belt situated between the orbits of Mars and Jupiter.

Astronomical unit A unit of distance defined by the mean distance of the Earth from the Sun and equal to 149,597,870 km.

Azimuth The angular distance along the horizon, measured in an eastward direction, between a point due north and the point at which a vertical line through a celestial object meets the horizon. (This is the normal convention for an observer in the northern hemisphere; other conventions are sometimes followed.) *See also* ALTITUDE.

Barycenter The center of gravity of a system of massive bodies; the barycenter of the Earth–Moon system, for example, lies at a point within the Earth's globe.

Black body An idealized body which reflects none of the radiation falling on it. Such a body would be a perfect absorber of radiation, and would emit a SPECTRUM determined solely by its temperature.

Bode's Law A curious numerical relationship between the distances of the various planets from the Sun. The law is often expressed in the form:
$$r_n = 0.4 + 0.3 \times 2^n,$$
where r_n is the distance of the planet from the Sun and n is $-\infty$ 0, 1, 2, 3 . . . in turn. The resulting values correspond surprisingly closely with the actual distances, but most astronomers consider this to be merely a coincidence.

Celestial equator The circle formed by the projection of the Earth's equator onto the surface of the CELESTIAL SPHERE.

Celestial sphere An imaginary sphere, centered on the Earth, onto whose surface the stars may be considered, for the purposes of positional measurement and calculation, to be fixed.

Chromosphere The layer of the Sun's atmosphere lying above the PHOTOSPHERE and below the CORONA.

Comet A type of heavenly body in orbit around the Sun, with several characteristics that distinguish it from the planets, satellites or asteroids. Comets typically have highly eccentric orbits, and some of them become bright objects in the sky as they approach PERIHELION, sometimes with a distinctive "tail". They are made up of a "nucleus" with a surrounding cloud of dust and gas which forms the "coma".

Conjunction The near or exact alignment of two astronomical bodies in the sky. Also used to describe an alignment between a planet and the Sun as seen from Earth. When the planet passes behind the Sun, the conjunction is called "superior"; in the special case of Mercury or Venus passing between the Sun and the Earth, the conjunction is called "inferior".

Coriolis effect The apparent deflection of a body moving in a rotating coordinate system. For example, a projectile fired northward from the Earth's equator will appear to be deflected to the east, because the point on the equator from which it is fired will be rotating faster than its target to the north. The Coriolis effect plays an important part in determining the directions of wind and ocean currents.

Corona The outermost part of the Sun's atmosphere. It is visible to the naked eye only during a total eclipse of the Sun, when it has the appearance of a halo around the Sun's obscured disc. The corona is the source of the SOLAR WIND.

Cosmic rays Extremely energetic atomic particles, principally protons, travelling through space at speeds approaching the speed of light. A proportion of cosmic rays come from the Sun, while the rest originate somewhere outside the SOLAR SYSTEM, possibly in violent events in the GALAXY.

Culmination The maximum altitude of a celestial body above the horizon.

Declination The angular distance of a celestial body from the CELESTIAL EQUATOR; one of the two celestial coordinates, roughly equivalent to latitude on the Earth, used to represent the position of a celestial object. *See also* RIGHT ASCENSION.

Doppler effect The apparent shift in the frequency of waves that occurs when there is relative motion between the source and the observer. A receding source will appear to emit waves of longer wavelength (or lower frequency) than it would if it were stationary; with an approaching source the effect is reversed, and the wavelength appears to be shorter (higher frequency).

Eclipse The partial or total disappearance of a celestial body either behind a nonluminous body or into its shadow. A solar eclipse, for example, occurs when the Sun is obscured by the Moon's disc, while a lunar eclipse takes place when the Moon passes through the cone of shadow cast by the Earth.

Ecliptic The circle on the CELESTIAL SPHERE defined by the Sun's apparent annual motion against the stellar background. The ecliptic represents the plane in which the Earth orbits the Sun and, because the Earth's rotational axis is tilted, the ecliptic is inclined to the celestial equator at an angle, known as the "obliquity of the ecliptic", which is equal to about $23\frac{1}{2}°$.

Electromagnetic radiation Radiation in the form of waves associated with electric and magnetic disturbances, which may be manifested in a variety of forms, such as light, X-rays and radio waves, depending on the wavelength. The electric and magnetic components are often represented as two waves oscillating in different planes at right angles to one another.

Elongation The angular distance of a planet from the Sun, or of a satellite from its primary planet.

Equation of time The difference between the apparent solar time and the mean time; the value of the equation of time varies throughout the year from about $-14\frac{1}{4}$ min to about $+16\frac{1}{4}$ min.

Exosphere The outermost region of the Earth's atmosphere, beyond the IONOSPHERE.

Faculae Bright patches on the PHOTOSPHERE of the Sun, normally associated with SUNSPOT groups.

First Point of Aries *See* VERNAL EQUINOX.

Flares Sudden brilliant outbursts in the outer part of the Sun's atmosphere, typically lasting only a few minutes. Generally associated with SUNSPOTS, they give rise to a type of COSMIC RAYS.

Fraunhofer lines Dark lines appearing in the spectrum of the Sun, resulting from the absorption of certain wavelengths of light by elements in the outer parts of the Sun's atmosphere.

Galaxy A large system of stars. The term "The Galaxy" refers to the particular galaxy of which the Sun is a member.

Hertzsprung-Russell diagram A graph on which is plotted the LUMINOSITY of stars against their temperature or spectral type. The diagram reveals that for a given spectral type, temperature is not randomly distributed. For the most numerous group of stars (the so-called "main-sequence" stars) the higher the temperature, the brighter, in general, is the star; other groupings in the H-R diagram represent stellar types such as Red Giants and White Dwarfs which do not obey this general rule.

Ion An atom that is electrically charged as a result of having lost or gained one or more electrons.

Ionosphere The region of the Earth's atmosphere, extending from approximately 80 km to 500 km above the surface, in which radiation from the Sun ionizes a substantial proportion of air molecules. *See also* ION.

Librations Apparent oscillations of the Moon as a result of which an Earth-based observer can see the surface from a slightly different angle at different times. Over a period of time, a total of about 59 percent of the Moon's surface can be seen from Earth.

Light-year A unit of distance defined by the distance travelled by light *in vacuo* in a year, equal to 9.4607×10^{12} km or 63,240 ASTRONOMICAL UNITS. In astronomy the more commonly used unit for large distances is the PARSEC, which is equal to 3.2616 light-years.

Limb The edge of the visible disc of a celestial body.

Luminosity The total amount of energy emitted by a star per unit of time.

Magnetosphere The region around a planet within which its magnetic field predominates over the magnetic field of the surrounding interplanetary region.

Magnitude A measure of the brightness of a star or other celestial body on a numerical scale which decreases as the brightness increases. The faintest stars visible to the naked eye on a clear night are of magnitude 6; the brightest have a mean magnitude of 1. The "absolute" magnitude of a star is defined as the apparent magnitude it would have if viewed from a standard distance of 10 PARSECS.

Meridian A great circle passing through the poles either of the Earth or of the CELESTIAL SPHERE. In astronomical usage, the term usually refers to the "observer's meridian", which passes through the observer's ZENITH.

Meteor A small particle of interplanetary material that leaves a bright trail across the sky as it burns up on entering the Earth's atmosphere.

Meteorite The remains of a METEOR that reaches the surface.

Meteoroid A small lump of solid meteoritic material in space.

Nodes The points at which two great circles on the CELESTIAL SPHERE intersect; in particular, the points at which the orbit of a body, such as a planet or the Moon, crosses the ECLIPTIC.

Occultation The temporary disappearance of one celestial body, usually a star, behind another, usually a planet or moon. A solar eclipse is a particular case of an occultation.

Opposition The position of a planet in its orbit when the Earth lies on a direct line between the planet and the Sun. A planet is best placed for observation when it is at opposition.

Parallax The apparent change in the position of an object due to an actual change in the position of the observer. Measurement of parallax allows the distances of distant objects to be determined.

Parsec A large unit of distance defined as the distance at which a star would have an annual PARALLAX of one second of arc, and equal to 3.0857×10^{13} km, 206,265 ASTRONOMICAL UNITS, or 3.2616 LIGHT-YEARS.

Penumbra The region of partial shadow that is formed around the region of total shadow when the source of illumination is of finite size. See also UMBRA. The term is also used to describe the outer part of SUNSPOTS.

Perihelion The point or moment of closest approach to the Sun of an orbiting body such as a planet. The opposite of APHELION.

Perturbations Irregularities in the orbital motion of a body due to the gravitational influence of other orbiting bodies.

Phase angle The angle defined by the position of the Sun, a body, and the Earth, measured at the body.

Photosphere The intensely luminous layer of the Sun that forms its visible surface.

Plage Bright areas of the PHOTOSPHERE associated with active areas on the Sun, caused by the presence of gas considerably hotter than its surroundings.

Planet One of the nine medium-sized bodies (including the Earth) which orbit the Sun; a similar body orbiting any other star. Unlike stars, planets do not emit their own heat or light from thermonuclear reactions in their interiors. The word "planet" is derived from a Greek word meaning "wanderer": the planets are seen to move against the background of fixed stars. An "inferior" planet is one whose orbit lies within that of the Earth, while a "superior" planet moves outside the Earth's orbit.

Polarization A special condition of ELECTROMAGNETIC RADIATION. Radiation (such as light) may be resolved into two components, one electrical, the other magnetic, at right angles to one another. When the radiation is unpolarized, the components vibrate in every direction, but if the radiation is "plane polarized", all the electrical components are arranged in planes parallel to each other, with their associated magnetic components lying at right angles to them. Other types of polarization, such as "circular" and "elliptical", are also possible.

Quadrature The position of the Moon or an outer planet when its ELONGATION is 90°.

Right ascension (R.A.) The angle, measured eastward along the CELESTIAL EQUATOR in units of hours, minutes and seconds, between the VERNAL EQUINOX and the point at which the MERIDIAN through a celestial object intersects the celestial equator. Right ascension is roughly equivalent to longitude on the Earth, and in conjunction with one other coordinate, DECLINATION, specifies the exact position of an object in the sky.

Roche limit The critical distance from the center of a planet within which gravitational forces would be insufficient to prevent a satellite from being broken up by tidal forces. For a satellite with the same density as the parent planet, the Roche limit lies at 2.4 times the radius of the planet.

Saros An interval of 6,583 days (equal to 18 years 11.3 days) after which the Sun, the Moon and the Earth return almost exactly to their previous relative positions. Consequently, the Saros period marks the interval between successive ECLIPSES of similar type and circumstance.

Sidereal period The time taken for a body to complete one orbit, as measured against the background of fixed stars. See also SYNODIC PERIOD.

Sidereal time A system of measurement of time based on the Earth's period of rotation, measured against the background of fixed stars. The sidereal day is taken to begin at the moment at which the VERNAL EQUINOX crosses the observer's MERIDIAN.

Solar constant The amount of energy per second that would be received in the form of solar radiation over one square meter of the Earth's surface at the Earth's mean distance from the Sun, if no radiation was absorbed by the atmosphere.

Solar cycle The periodic variation of solar activity, as manifested in the number of SUNSPOTS, the frequency of solar FLARES and various other solar phenomena. The cycle has an average period of about 11 years.

Solar System The system made up of the Sun, the planets (Mercury, Venus, Earth, Mars, Jupiter, Saturn, Uranus, Neptune and Pluto) together with their satellites, the ASTEROIDS, COMETS, METEOROIDS and interplanetary material.

Solar wind An electrically charged stream of atomic particles, mainly protons and electrons, emitted by the Sun.

Solstices The two points on the ecliptic of maximum or minimum DECLINATION; the times at which the Sun reaches these points along its annual path. The summer solstice (corresponding to the maximum declination) falls around 21 June, the winter solstice (minimum declination) around 21 December.

Spectrum The range of wavelengths or frequencies present in a sample of ELECTROMAGNETIC RADIATION. Visible radiation (i.e. light) may be resolved into its component wavelengths by passing it through a prism; white light will be spread out into a band of colors. A glowing gas under low pressure will emit radiation only at certain specific wavelengths, which appear as bright, isolated "emission" lines in its spectrum; similarly, it will only absorb radiation at these same wavelengths. When radiation is absorbed from a "continuous" spectrum, black "absorption" lines appear. See also FRAUNHOFER LINES.

Stratosphere The region of the Earth's atmosphere, extending from about 15 km to 50 km above the Earth's surface, between the TROPOSPHERE and the mesosphere.

Sunspots Large transient patches on the PHOTOSPHERE of the Sun which appear black in contrast with the surrounding regions. The number of sunspots varies in a periodic way (see SOLAR CYCLE).

Synchrotron radiation Radiation emitted by electrons travelling in a strong magnetic field at speeds approaching the speed of light.

Synodic period The interval between successive CONJUNCTIONS or, more generally, between similar configurations of a celestial body, the Sun and the Earth.

Tektites Small glassy objects, found in a few restricted areas of the Earth, whose origin remains a mystery; believed to be associated with METEORITE impacts on Earth.

Terminator The boundary between the dark and the sunlit hemispheres of a planet or satellite.

Troposphere The lowest layer of the Earth's atmosphere, within which temperature decreases with increasing altitude. It extends to a height of about 15 km.

Tropopause The boundary between the TROPOSPHERE and the STRATOSPHERE.

Umbra The dark central region of a shadow. See also PENUMBRA.

Vernal equinox The point on the CELESTIAL SPHERE at which the ECLIPTIC crosses the CELESTIAL EQUATOR from south to north (where the direction is defined by the Sun's motion). Also known as the First Point of Aries.

Zeeman effect The splitting of spectral lines (see SPECTRUM) when emission or absorption occurs in the presence of a strong magnetic field.

Zenith The point on the CELESTIAL SPHERE directly above the observer.

Zodiac A belt on the CELESTIAL SPHERE extending by about 8° on either side of the ECLIPTIC, marking the region within which the Sun and the planets are always to be found. The zodiac is divided into 12 equal zones which are named after 12 constellations.

Lunar Probes 1958-1980

Name	Country	Launch d m yr	Remarks
Thor-Able 1	USA	17. 8.58	First attempt to reach the Moon. Failed.
Pioneer 1	USA	11.10.58	Failed. Sent data for 43 hours.
Pioneer 2	USA	8.11.58	Failed to reach the Moon.
Pioneer 3	USA	6.12.58	Failed, but provided radiation data.
Luna 1	USSR	2. 1.59	Passed Moon at 6,000 km. Went into solar orbit.
Pioneer 4	USA	3. 3.59	Passed Moon at 60,000 km. Went into solar orbit.
Luna 2	USSR	12. 9.59	First probe to hit the Moon, crash landing at 30°N, 1°W.
Luna 3	USSR	4.10.59	Photographed the Moon's far side.
Atlas-Able 4	USA	26.11.59	Failed to reach the Moon.
Atlas-Able 5	USA	25. 9.60	Failed to reach the Moon.
Atlas-Able 5B	USA	15.12.60	Failed to reach the Moon.
Ranger 3	USA	26. 1.62	Missed Moon by 36,800 km.
Ranger 4	USA	23. 4.62	Hit the Moon (at 15°5S, 130°7W), but cameras failed.
Ranger 5	USA	18.10.62	Missed Moon by 725 km. In solar orbit.
Unnamed	USSR	4. 1.63	Probable unsuccessful lunar probe.
Luna 4	USSR	2. 4.63	Unsuccessful soft-lander. Missed the Moon by 8,500 km, and entered solar orbit.
Ranger 6	USA	30. 1.64	Hit the Moon (at 0°2N, 21°5E), but TV system failed; no data.
Ranger 7	USA	28. 7.64	Landed in Mare Nubium at 10°7S, 20°7W. Returned 4,308 photographs.
Ranger 8	USA	17. 2.65	Landed in Mare Tranquillitatis at 2°7N, 24°8E. Returned 7,137 photographs.
Cosmos 60	USSR	12. 3.65	Probable unsuccessful lunar probe.
Ranger 9	USA	21. 3.65	Landed in Alphonsus (12°9S, 2°4W). Returned 5,814 photographs.
Luna 5	USSR	9. 5.65	Crashed in Mare Nubium at 31°S, 8°E; failed soft-lander.
Luna 6	USSR	8. 6.65	Missed Moon by 161,000 km (11 June) and entered solar orbit.
Zond 3	USSR	18. 7.65	Passed Moon at 9,219 km; returned 25 pictures of the far side. Entered solar orbit.
Luna 7	USSR	4.10.65	Crashed in Oceanus Procellarum at 9°N, 40°W. Failed soft-lander.
Luna 8	USSR	3.12.65	Crashed in Oceanus Procellarum at 9°1N, 63°3W. Failed soft-lander.
Luna 9	USSR	31. 1.66	Successful soft-lander in Oceanus Procellarum at 7°1N, 64°4W. 100 kg capsule landed. Photographs returned.
Cosmos III	USSR	1. 3.66	Probable unsuccessful lunar probe.
Luna 10	USSR	31. 3.66	Lunar satellite; minimum distance from Moon 350 km; contact maintained for 460 orbits in two months.
Surveyor 1	USA	30. 5.66	Landed at 2°5S, 43°2W, near Flamsteed; returned 11,237 photographs.
Explorer 33	USA	1. 7.66	Failed lunar orbiter.
Lunar Orbiter 1	USA	10. 8.66	Photographed Moon until 29 August 1966. Impacted at 6°7N, 162°E, on 29 October 1966.
Luna 11	USSR	24. 8.66	Minimum distance 159 km. Transmitted until 1 October 1966.
Surveyor 2	USA	20. 9.66	Unsuccessful soft-lander; crashed at 5°N, 25°W, near Copernicus.
Luna 12	USSR	22.10.66	Transmitted until 19 January 1967.
Lunar Orbiter 2	USA	6.11.66	Lunar satellite. Transmitted 422 pictures before impacting at 4°S, 98°E.
Luna 13	USSR	21.12.66	Soft-landed at 18°9N, 62°W in Oceanus Procellarum. Transmitted until 27 December 1966. Soil studies.
Lunar Orbiter 3	USA	4. 2.67	Lunar satellite, impacted at 14°6N, 91°7W. Returned 307 pictures.
Surveyor 3	USA	17. 4.67	Landed at 2°9S, 23°3W in Oceanus Procellarum, 612 km east of Surveyor 1, near Apollo 12 site. Returned 6,315 pictures. Soil physics.
Lunar Orbiter 4	USA	4. 5.67	Returned 326 pictures.
Surveyor 4	USA	14. 7.67	Failed. Crashed at 0°4N, 1°3W in Sinus Medii.
Explorer 35	USA	19. 7.67	Studies of Earth's magnetic field.
Lunar Orbiter 5	USA	1. 8.67	Lunar satellite. Controlled impact at 0°, 70°W on 31 January 1968.
Surveyor 5	USA	8. 9.67	Landed at 1°4N, 23°2E in Mare Tranquillitatis, 25 km from Apollo 11 site. Returned 18,006 pictures.
Surveyor 6	USA	7.11.67	Landed at 0°5N, 1°4W in Sinus Medii. Returned 30,065 pictures.
Surveyor 7	USA	7. 1.68	Landed at 40°9S, 11°5W on north rim of Tycho. Returned 21,274 pictures. Soil analyses.
Zond 4	USSR	2. 3.68	Lunar probe, but exact purpose unknown.
Apollo 6	USA	4. 4.68	Failed to reach the Moon.
Luna 14	USSR	7. 4.68	Lunar satellite; minimum distance from Moon 160 km.
Zond 5	USSR	15. 9.68	Went round Moon and landed in Indian Ocean, 21 September 1968.
Zond 6	USSR	10.11.68	Went round Moon and returned to Earth on 17 November 1968.
Apollo 8	USA	21.12.68	Manned orbiter; 10 orbits completed. (Lovell, Borman, Anders.)
Apollo 10	USA	18. 5.69	Manned orbiter; went to within 14.9 km of the Moon and tested Lunar Module. (Stafford, Cernan, Young.)
Luna 15	USSR	13. 7.69	52 orbits; crashed at 17°N, 60°E in Mare Crisium, 21 July 1969.
Apollo 11	USA	16. 7.69	Manned landing at 0°7N, 23°4E, 20 July 1969, in Mare Tranquillitatis. (Armstrong, Aldrin, Collins.)
Zond 7	USSR	8. 8.69	Went round Moon and returned to Earth.
Apollo 12	USA	14.11.69	Landed on 19 November 1969 at 3°2S, 23°8W in Oceanus Procellarum. (Conrad, Bean, Gordon.)
Apollo 13	USA	11. 4.70	Unsuccessful manned lander; returned 17 April 1970. (Lovell, Haise, Swigert.)
Luna 16	USSR	12. 9.70	Landed at 0°7S, 55°3E in Mare Fecunditatis, 0°41S, 56°18E, on 20 September 1970. Returned 100 g of soil.
Zond 8	USSR	20. 9.70	Orbited Moon; returned 27 October 1970
Luna 17	USSR	10.11.70	Carried Lunokhod 1 to Moon; landed at 38°3N, 35°W in Mare Imbrium. 17 November 1971, returning over 20,000 pictures.
Apollo 14	USA	31. 1.71	Manned lander. Landed at 3°7S, 17°5W in Fra Mauro on 5 February 1971. (Shepard, Mitchell, Rossa.)
Apollo 15	USA	26. 7.71	Manned lander; Hadley–Apennines (26°1N, 3°7E). (Scott, Irwin, Worden.)
Luna 18	USSR	2. 9.71	Crashed at 3°6N, 56°5E in Mare Fecunditatis after 54 orbits.
Luna 19	USSR	28. 9.71	Contact maintained for over a year and 4,000 orbits.
Luna 20	USSR	14. 2.72	Landed 21 February 1972 at 3°5N, 56°6E in Mare Fecunditatis 120 km from Luna 16's impact. Returned samples; landed on Earth 25 February 1972.
Apollo 16	USA	16. 4.72	Landed at 8°6S, 15°5E in Descartes area on 21 April 1972. (Young, Duke, Mattingley.)
Apollo 17	USA	7.12.72	Landed at 21°2N, 30°6E in Taurus–Littrow, 11 December 1972. (Cernan, Schmitt, Evans.)
Luna 21	USSR	8. 1.73	Carried Lunokhod 2 to Lemonnier area on 16 January 1973. Lunokhod 3 transmitted until 3 June 1973, returning over 80,000 pictures.
Explorer 49	USA	10. 6.73	Radio astronomy from far side.
Luna 22	USSR	29. 5.74	Transmitted until 6 November 1975.
Luna 23	USSR	28.10.74	Landed in Mare Crisium. Sampling unsuccessful. Transmitted until 9 November 1975.
Luna 24	USSR	9. 8.76	Landed at 12°8N, 62°2E in Mare Crisium. Drilled to 2 m. Landed back on Earth 22 August 1976.

Observing the Moon

No valuable research can be carried out until the observer is thoroughly familiar with the Moon in all its aspects. Photographic studies, which require an equatorial mount and clock drive, are helpful even though they have little scientific value. For the sake of gaining experience the observer should also make numerous drawings of features from direct telescopic observations. It is unwise to use too small a scale in a lunar sketch; generally, 1 cm to 12 km is suitable: Plato, for example, would be 8 cm in diameter.

For familiarization a small telescope (15 cm reflector) is adequate, but for serious research a larger aperture – at least 19 cm is required. Much, however, depends upon the observer's skill.

Regarding magnification, it is generally held that a telescope will bear a power of 25 per cm of aperture under good conditions. It is important not to use too high a magnification: a small, sharp image is preferable to a larger, slightly blurred one.

Transient Lunar Phenomena

One of the most important fields of research today is that of TLP (*see* page 21). There is very strong evidence in favor of their reality (quite apart from N. A. Kozyrev's classic observation of a red event in Alphonsus, in 1958), but knowledge of them is still very incomplete, and although many theories have been put forward the cause and distribution of TLP are still not understood with certainty. Some TLP are reddish, but the color is never striking. One method of detection involves using a "Moon-Blink" device, which consists of a rotating filter, half red and half blue. The suspected formation is viewed through each filter in quick succession, and any red area will then show up as a "winking patch". Yet it is only too easy to be misled. If a TLP is suspected, the first step is to examine other formations in the same area. If these too show the same phenomenon, then it is clear that conditions in the Earth's atmosphere are responsible. Other TLP take the form of localized obscurations, such as a slight blurring of detail over a restricted area. Here again, features near the suspected formation should be examined to make sure that the effect is not due to the terrestrial atmosphere.

There are some features, such as Aristarchus, Alphonsus and Gassendi, that are believed to be particularly prone to TLP, and these should be surveyed whenever possible; but it is also important to survey regions where no TLP have ever been reported.

If a TLP is suspected, the following details should be noted: date, time (GMT), telescope aperture, magnification and seeing conditions using the Antoniadi scale from 1 (perfect) to 5 (poor). (Generally, no observation is to be regarded as even moderately reliable unless the seeing is at least 3.) The area concerned should be carefully noted, and either sketched or (preferably) outlined on a photograph. (The conditions of lighting and libration in the photograph will not be the same as that during the observation itself.)

Confirmation from an independent observer at a different site is needed. Most countries have astronomical societies coordinating this work, such as the British Astronomical Association in the U.K. and the Association of Lunar and Planetary Observers in the U.S. The program involves many hours of fruitless searching, and the very existence of observable TLP is still doubted by some authorities. Apart from Kozyrev's original observation of 1958 there is still a lack of spectroscopic confirmation.

Finally, observations of occultations are still required. They can be timed to within an accuracy of 0.1 sec, and these too are coordinated by national societies. Occasionally stars appear to fade out instead of disappearing instantaneously; it will generally be found that in such cases the star is a close double.

Lunar observation is still worth undertaking. The main difference between the observer of the 1980s and his predecessor of the 1940s is that the modern lunar researcher has to be much more specialized in his approach.

Bibliography

Books

Adams, P., *Moon, Mars and meteorites* (HMSO, 1977)

Bowker, D. E. and Hughes, J. K., *Lunar Orbiter Photographic Atlas of the Moon* (NASA SP-206, 1971)

French, B., *The Moon Book* (Penguin Books, 1977)

Guest, J. E., and Greeley, R., *Geology on the Moon* (Wykeham publications (London) Ltd., 1977)

Gutschewski G. L. *et al*, *Atlas and Gazetteer of the Near Side of the Moon* (NASA SP-241, 1971)

Ibid., On the Moon with Apollo 16; A Guidebook to the Descartes Region (NASA, 1972)

Kosofsky, L. J. and El-Baz, F., *The Moon as Viewed by Lunar Orbiter* (NASA SP-200, 1970)

Lindsay, J. F., *Lunar Stratigraphy and Sedimentology; Developments in Solar System and Space Science-3* (Elsevier, 1976)

Levinson, A. A. and Taylor, S. R., *Moon Rocks and Minerals* (Pergamon Press, 1971)

Lewis, H. A. G., ed., *The Times Atlas of the Moon* (Times Newspapers Ltd., 1969)

Mason, B. and Melson, W. G., *The Lunar Rocks* (Wiley Interscience, 1970)

Masursky, H. *et al*, ed., *Apollo over the Moon; a View from Orbit* (NASA SP-362, 1978)

Moore, P., *Guide to the Moon* (Lutterworth Press, 1976)

Norton, A. P., *Norton's Star Atlas* (Gall and Inglis Ltd., 1978)

Proceedings of the Lunar and Planetary Science Conferences **1–11** (Pergamon Press, 1979–81)

Ringwood, A. E., *Origin of the Earth and Moon* (Springer Verlag, 1979)

Schultz, P. H., *Moon Morphology* (University of Texas Press, 1972)

Short, N. M., *Planetary Geology* (Prentice-Hall, Inc., 1975)

Simmons, G., *On the Moon with Apollo 15; A Guidebook to Hadley Rille and the Apennine Mountains* (NASA, 1971)

Taylor, S. R., *Lunar Science: A Post-Apollo View* (Pergamon Press Inc. 1975)

The Handbook of the British Astronomical Association, 1980

Articles

Wood, J. A., "Origin of Earth's Moon", in *Planetary Satellites* (ed. Burns), 513–529 (1977)

Wood, J. A., "The Moon", *Scientific American*, *233*, 93–102 (1975)

Weaver, K. F., "The Moon", *National Geographic*, *135*, ii, 207–239, (1969)

Index

Figures in Roman type refer to text entries; figures in *italic* refer to illustrations or captions.

Photographic Credits
p.5(1) Lick Observatory
p.5(2) Staatliche Museen Berlin, Antikenabteilung
pp.8–9(2B, 3rd from left) Observatoire de Paris
pp.8–9(2B, all others) Lick Observatory
pp.10–11(1) A.H. Mikesell
p.12(1) Bibliothèque Nationale, Paris/Photo: Jean-Loup Charmet
p.12(2) A.R. Michaelis Collection
p.12(3) Library of the Strahov Monastery, Prague
p.13(4, 5, 6) Bibliothèque Nationale, Paris/Photo: Jean-Loup Charmet
p.22(3B) Institute of Physics, Bristol
p.23(5B) Courtesy Dr. M. Maurette & Dr. J.-C. Dran
p.23(5C) Courtesy Prof. R.M. Walker, University of Washington
p.23(5D) Pergamon Press
p.25(6A, 6B, 6C) Don E. Wilhelms & Donald E. Davis, U.S. Geological Survey

p.31(4A, 4D) D. Roddy, U.S. Geological Survey
p.31(4B) West Australia Department of Tourism
pp.58–9, 64–5, 70–1 Lunar and Planetary Laboratory, University of Arizona
Patrick Moore Collection: p.13(7), p.31(4C)
NASA: p.43 (bottom); NASA/Lyndon B. Johnson Space Center: p.22 (2A, 2B, 2C, 2D, 3A), p.23 (4A, 4B, 4C, 4D, 5A), p.28(3), p.48 (top, bottom); NASA/Jet Propulsion Laboratory: p.31 (3); NASA/Picturepoint: p.33, 34, 47; NASA/Woodmansterne Ltd: p.35, 36, 37, 40, 43 (top)
Apollo and Lunar Orbiter photography provided by the National Space Science Data Center through the World Data Center A for Rockets and Satellites: p.22 (1), p.27 (3A, 3B), p.29 (4), p.32 (1–6), pp.38–9, p.41, p.42, pp.44–5, p.46, pp.74–5, pp.76–7, pp.80–1, pp.82–3, pp.86–7

Illustrators
Kai Choi, Chris Forsey, Mick Saunders, Charlotte Styles